W9-BLU-287

TEC

Wind Energy
Technicians

Mike Jones **DISCARD**

TSTC
Publishing

© 2010 TSTC Publishing

ISBN 978-1-934302-55-2

TSTC Publishing
Texas State Technical College Waco
3801 Campus Drive
Waco, Texas 76705
http://publishing.tstc.edu/

Publisher: Mark Long
Project manager: Grace Arsiaga
Marketing: Sheila Boggess
Sales: Wes Lowe
Printing production: Data Reproductions Corporation
Indexing: Michelle Graye, indexing@yahoo.com
Book design & layout: James Brown
Editor: Joseph Abbott
Cover design: The Cowley Group
Special thanks to Cloud County Community College for the photographs found on pages 3, 10, 32, 47, 60, 87, and 88.

Manufactured in the United States of America
First edition

Publisher's Cataloging-in-Publication
(Provided by Quality Books, Inc.)

Jones, Mike, 1946 Dec 20-
    Wind energy technicians / Mike Jones. -- 1st ed.
    p. cm. -- (TechCareers)
    Includes index.
    ISBN-13: 978-1-9-34302-55-2
    ISBN-10: 1-934302-55-4

    1. Wind power--Vocational guidance.
    2. Industrial technicians--Vocational guidance.
    I. Title.
    II. Series: TechCareers.

TJ820.J66 2010   621.4'5'023
                           QBI10-600052

Table of Contents

Commonly Used Abbreviations

AAS	Associate of Applied Science degree
ASC	Applied Science Certificate
AWEA	American Wind Energy Association
BCE	Before the Common Era
EHS	Environmental Health and Safety
GW	Gigawatt (1,000,000,000 watts)
IEC	International Electrotechnical Commission, the recognized international body for standards development activities
MSCHE	Middle States Commission on Higher Education
MW	Megawatt (1,000,000 watts)
NEASC	New England Association of Schools and Colleges
NCA	North Central Association of Colleges and Schools
NWCCU	Northwest Commission on Colleges and Universities
OAPEC	Organization of Arab Petroleum Exporting Countries (consisting of the Arab members of OPEC plus Egypt and Syria)
OEM	original equipment manufacturer

OSHA	Occupational Safety and Health Administration
PLC	programmable logic controller
PPE	personal protective equipment
rpm	revolutions per minute
SACS	Southern Association of Colleges and Schools
SCADA	Supervisory Control And Data Acquisition
TC-88	Technical Committee-88 of the International Electrotechnical Commission, the recognized international body for standards development activities.
WASC	Western Association of Schools and Colleges
WET	wind energy technician
WIE	Women in Engineering
WoWE	Women of Wind Energy

CHAPTER ONE

Wind Energy Technician Careers

Every time you turn on the lights in your house, open a refrigerator that keeps your food safely cold, watch a movie on your flat-screen TV, charge up a cell phone, or begin to play on a gaming console, you may not give much thought to how the power was generated that ultimately made its way to you. After all, most people see energy as being much like oxygen: life sustaining, all around us, and yet, at the same time, fundamentally invisible.

Then again, you could be one of those rare breed of people who, with forty pounds of tools and gear in tow, will climb up 300 feet inside a narrow tower, pull themselves into a small compartment full of rotors and gears, throw open the top hatch, and climb out for a bird's eye view of the world that few people will ever see. Right in front of you is the potential for raw power generation that even fewer people will ever see this close and personal: the carefully engineered blades of a wind turbine.

Since the Arab oil embargo in the early 1970s, both multinational corporations and the governments of industrialized nations have begun to develop alternative energy sources for the 21st century and beyond through solar, geothermal, hydrogen-based fuel-cell systems, and nuclear power. In the last ten years, wind energy has become increasingly popular as a form of energy production, especially with its smaller environmental footprint in comparison to its counterparts such as nuclear power.

So, with world oil reserves steadily decreasing, opportunities for careers related to any of these alternative energy solutions are already so numerous that qualified technicians are, and will likely remain, in great demand. "Wind farms" are now a more and more familiar sight all across Texas, the Midwest, and coastal areas of the United States, as well as in other parts of the globe. The technical and interpersonal skills, as well as education needed to succeed as a certified wind energy technician, will be transferable to virtually any location in the world.

Wind Energy Technology Overview

Wind energy is one of humankind's oldest tools, with various applications originating independently in numerous parts of the world throughout ancient history. For more than 5,500 years people have been using wind power to propel sailing vessels and to provide natural ventilation. Harnessing wind energy for mechanical power dates from the seventeenth century BCE when the Babylonian emperor Hammurabi designed wind energy systems to power massive irrigation projects. The first simple vertical-axle windmills appeared in Afghanistan in the seventh century of the Common Era and were used to grind grains and draw well water. Horizontal-axle windmills appeared in northern Europe in the twelth century, and many of these historic windmills are still a common sight in Belgium and the Netherlands.

The water-pumping windmill was one of the central factors in the successful development of farming and ranching on arid grasslands which did not provide easily accessible water in much of North America. Windmills were also integral to the expansion of railroads in the nineteenth century, providing steam locomotives with a plentiful and dependable supply of water along transport routes.

In the late 1880s, American and British inventors simultaneously introduced wind turbines designed to provide electric lighting for businesses and homes. The first modern wind turbines were built in the early 1980s, although more efficient designs are still being developed by industry leaders like Vestas (Denmark), GE Energy (US), Gamesa and Acciona

(Spain), Enercon, Siemens and Nordex (Germany), Goldwind and Sinovel (People's Republic of China) and Suzlon (India).

At the end of 2007, wind-powered generators worldwide generated approximately 94.1 gigawatts (GW) annually. Although only about one percent of world-wide electricity is provided by wind energy, it is a rapidly growing resource, increasing by more than 500 percent globally between 2000 and 2007. Several western European nations have already integrated wind energy sources for their electrical requirements, including Denmark (nineteen percent, Spain and Portugal (nine percent), Germany (six percent) and the Republic of Ireland (six percent).

Although more popular overseas, electricity generated from wind turbines is increasingly in demand across the United States and Canada. State and federal government incentives, rapidly improving technology, and renewable energy market credits are driving today's wind power industry. The American Wind Energy Association (AWEA) reports that the American wind industry grew by forty-five percent in

2007. Individuals with education and training in electrical and mechanical engineering, meteorological and earth sciences, and project management are finding more and more lucrative opportunities in the wind energy industry.

Many environmentalists favor renewable or "green" power resources such as wind, solar, hydroelectric, tidal and geothermal as alternatives to fossil fuels like coal and petroleum. These alternate energy sources are considered plentiful, unlimited, easily distributed and clean—with lower greenhouse gas emissions. Wind energy, in the form of widespread wind farms or interconnected wind turbines, is still controversial because of its visual impact on the environment and what some see as its land use complications.

What is a Wind Turbine?

A wind turbine is a rotating machine that enables conversion of the wind's kinetic energy into mechanical energy. A wind farm is merely a large collection of wind turbines in the same location, all of them usually connected to the local electric power grid via power collection system and communications network. Smaller turbines or groups are often used to provide electricity to isolated locations.

A large wind farm might consist of a few dozen to about 100 individual wind turbines, and cover an extended area of hundreds of square miles, but the land between the turbines can also be used for agricultural or other purposes. Some wind farms are located offshore to take advantage of strong winds blowing over the surface of a large body of water.

The intermittent nature of wind power does not seem to seriously limit its usefulness in the overall context of supply and demand, and additional costs due to wind intermittency are considered modest. This benefits electricity customers both at the regional grid level and at the community level because the price of wind power is now as cost-effective as other common energy sources like coal and natural gas. The result is the

dramatic increase in the number of wind farms appearing throughout Texas, the Southwest and the upper Midwest.

With increasing demand for wind power, there is already a pressing need for wind technicians to maintain the wind turbines. Regional programs to train new students in wind energy technology and economics, turbine maintenance, and tower construction, erection and tower safety are a growing necessity. Additional research is needed immediately for more research on developing technologies to make wind power a reliable resource, available on demand rather than just when the wind blows.

Employment Outlook

The ceaseless winds blowing across America are more than enough to supply everyone in our nation with electricity. The United States has taken advantage of this energy resource; from 2005-2008, U.S. businesses and industries have led the world in wind power generation. Hundreds of companies currently are dedicated strictly to the generation of wind energy, and more than 16,000 companies now employ more than one million people in manufacturing equipment and other products that support wind energy generation. For example, the Titan Wind Project is close to completion in South Dakota. The 2,000 turbine, 5,050 megawatt (MW) wind farm is capable of supplying energy to 1.5 million homes—more than five times the number of homes in the state! In neighboring North Dakota, FPL Energy's Oliver County Wind Energy Center opened recently with 54 turbines capable of powering more than 24,000 homes.

About twenty-five percent of the electrical power currently used in the greater Houston, Texas, area is now provided by wind energy. Soon after Hurricane Katrina severely damaged the oil and gas production infrastructure and increased the cost of natural gas, the Houston City Council took steps to provide cleaner, cheaper energy to its citizens via wind power.

Bluewater Wind, under construction off the coast of Delaware, is the first offshore wind farm in the U.S. and will be completed in 2012. The winds near Rehoboth Beach are capable of producing an average yearly output of 5,200 MW, or four times the average consumption of the state of Delaware. The financial and business potential for Bluewater Wind is enormous: It could produce more than $2 billion worth of yearly revenue on the wholesale electricity market, create hundreds of new jobs, and provide roughly $100 million in wages for local construction workers.

WET Career Profile: *Alice Orrell*

The Department of Defense is largest energy consumer in the United States, according to Brian Lally, deputy undersecretary of defense for installations and environment. An April 26, 2009, *Los Angeles Times* article also reported that the DOD now considers reducing energy consumption by turning to green alternatives to be a matter of homeland security.

Alice Orrell, an energy analyst at the Department of Energy's Pacific Northwest National Laboratory in Richland, Wash., makes it her mission to help the DOD comply with the mandates to reduce the military's carbon footprint and reach its goals of using a greater percentage of renewable energy on its bases.

Before Orrell began her work with the DOE, however, she earned a mechanical engineering degree from the University of Vermont, followed by a master of business administration from the University of Washington. Her career began in the power industry with the total opposite of wind energy: combustion engines. However, she had developed an interest in alternative energy throughout college and knew that was the direction she'd like her career to go. Once she had learned the power industry, she used earning her MBA, combined with her engineering degree and experience, to transition into the wind industry.

While finishing the work on her MBA, she gained an internship with a local non-profit organization that focused on wind energy. From that stepping stone, she moved to consulting from home when her husband took a job at the National Lab, requiring a family move. After two years of consulting, she attended a wind power conference.

"I started talking with a woman who worked at the National Lab," Orrell said. "She encouraged me to go for an interview. So I finally did, and one thing led to another. When a position opened, I got the job."

Although the National Lab performs a variety of scientific research, Orrell's group focuses on energy efficiency and renewable energy, using lighting standards and green building code standards. Orrell, who works part time in order to spend more time with her preschooler, works with one very specific client: the U.S. Army.

"I have a lot of different projects where our clients have goals and mandates that they have to meet on how much of their energy must come from renewable," Orrell said. "I look at their sites to assess whether wind turbines can be sited there. We've done our land assessments, and now we're in the process of putting up meteorological towers to measure wind. We're designing the sites and putting together a project layout so the clients can see how it will affect the land use. Being the army, they use large-scale equipment and conduct training maneuvers in these areas. We have to be sure the siting of the turbines is effective, not only for wind use but for their land use."

Orrell enjoys the daily interaction with her clients. "When working at a national lab, there's a lot of research. But my work is tangible, putting together towers and site projects, not just doing behind the scenes research. I like some of the actual implementation support that I give when consulting with my Department of Defense clients."

Orrell is also involved with the wind energy community in other ways, as well. A mentor in the Women of Wind Energy (WoWE) organization, Orrell said that like most engineering and technical fields, women are by far the minority. "Across the board," she said, "only about twenty percent of women are engineers. It's the same in the wind industry."

Although Orrell said she has not encountered any direct sexism or prejudice within the field, she believes it's important for women to network and provide support for one another. She's not only a mentor in WoWE, she heads up the mentoring program.

"I do program coordination for matching mentors with mentees, and I'll be working with a mentee soon," she added. "The goal is to use the program to set up an agreement plan for how often the pair should talk, and on confidentiality as often these women are in competing businesses. Then we form an action plan where the mentee has certain goals and the mentor helps her identify action items she can do to accomplish those goals."

This busy engineer, balancing career goals with motherhood, is an excellent example to anyone, but especially other women seeking to get into a wind energy career.

"We don't need more people like me, we need more people to climb the towers and keep the turbines working," she advises. "But, for where I work, it's necessary to have an engineering degree or you'll find the work more difficult. To do any kind of technical analysis, like how much energy will your project make, or where do the turbines need to be located, how should they connect to the grid, etc., it's almost mandatory to have an engineering degree. Definitely a college degree of some sort is required to work in the National Lab. This industry likes technical people.

"When I interviewed here, they asked what I considered to be my greatest accomplishment. I told them it was networking and developing a huge database of networks from going to conferences and getting involved in WoWE. So, I would say 'Tell potential employees to network. Get to know people in the career.' President Obama wants the US to have twenty percent of our total energy usage coming from renewable energy by 2030. That's driving everyone's work right now. It's a great time to start making connections."

Wind Energy Jobs and Career Paths

Wind power companies urgently are looking for technicians, engineers, and individuals with management experience in related fields. Wind energy companies often provide in-house training for new employees, along with educational opportunities for further licensing and certification. The many new formal education and training programs across the United States are a direct result of the immediate need for professionals with background in wind energy and are making the field attractive not only to newcomers, but to technicians and engineers in other fields.

Career search engines such as Technical Job Search (technicaljobsearch.com/), NationJob (www.nationjob. com/), and Wind Industry Jobs (www.windindustryjobs. com) provide an interesting glimpse into the types of jobs and salary ranges currently available in the wind and other "clean energy" industries, compared to those in traditional oil field careers. The following list is a brief, general representation of the types of jobs for which wind energy companies nationwide are searching for qualified candidates.

Wind Field Technicians

Trained wind field technicians provide day-to-day monitoring and maintenance of turbines and other machinery on the wind farm. Although a bachelor's degree is preferred, there is an increasing need for individuals with certification or an associate of applied science (AAS) degree in wind energy technology, as well as those with experience as an electrician or in other technical occupations. A wind turbine technician (WET) can often earn an average yearly salary of $36,000.

WET Career Profile: *Christopher Plummer*

For a single father charged with the responsibility of raising a small child, a career in wind energy can be life changing. And that's exactly what Christopher Plummer is expecting. "This renewable energy boom is just like the oil boom of the 1910s and '20s," he said. "It's the new energy revolution, and it's going to be here for awhile."

Plummer first became interested in the industry in about 2005, while he was working for Morgan Pools and Spas. His father got a job in wind industry, and the more Plummer heard his dad talk, the more his interest was piqued. So much so, that he left his job at Morgan to take a job at a fast food drive-in eatery, so he could have flexible hours to take online courses at Texas State Technical College. Considering the money Plummer expects to make upon graduation, he said he felt the scrimping and sacrifice was worth the potential pay-off down the road.

Working until midnight most days, and then doing homework until two a.m., finally falling into bed and getting up the next day to finish his homework before going back to work is a routine that would deter many from following their dreams; but not Plummer.

"My first online class was a real challenge," he admitted. "Just disciplining myself for homework again, to make time to get on the computer and attend class, was a challenge. But I love a challenge, and I fell right into the groove."

In fact, Plummer said he has been thrilled with his experience at TSTC and would recommend the school to anyone interested in the wind industry. A certificate level student, Plummer finished the online portion of his studies and took the six-week lab portion which included hands-on experience with many of the applications and theories he has learned, as well as the safety aspects of the job. Taking the safety section has also given him more confidence about the job conditions he will face.

"My instructors are very in-depth and love for you to ask questions," he said. "That's been a great help to me. We currently have a guy from Siemens who is auditing the class to see how Siemens can use it, and he's been actually sharing and teaching from his experience, so that is a great opportunity to talk to someone in the field."

Talking to people in the field is something Plummer has seen as important in preparing for his career. He said, "I talk to everyone I can. The people in the wind industry are very proud of what they're doing; they all love to talk about their jobs. I ask them what their favorite part is, what they actually do, etc. They are very happy to tell me all about it."

Talking to current wind energy employees has served to excite Plummer about his potential within the career. "I think it will be amazing," he said. "I can't imagine making that much money, because the opportunity is so huge for me. I might get a chance to do a little traveling. I'll be getting the chance for the first time in my life since being a single dad to buy a house, get a new vehicle, and be comfortable. I'm looking forward to that. No doubt, once I get in the door, this is going to financially change my life."

It's a mantra Plummer willing passes on to other people considering this career.

"This is a boom," he tells them. "Horse Hollow Wind Farm near Sweetwater is the biggest wind farm in the world now. The turbines out here came up seemingly overnight, and it's the same in a lot of other places. If you're thinking about getting into the career, do it now. Start studying; find everything you can on the subject. Start those classes. The way the industry is heading, you will need to have education to walk onto a [wind] farm and get a job. With the economy the way it is, a lot of people are out of jobs and are going to be looking. The ones with education will have a head start in this career."

Project Manager

Flexible individuals who enjoy interesting jobs that offer variety, travel, and opportunities for meeting new people are ideal candidates for positions as wind farm project manager. They may frequently visit various wind farms located throughout the United States (and some countries) to set up contracts and implement operations. Project managers supervise projects from inception to completion, working with contractors and providers of technical services until the project is completed and ready for implementation. Preferred candidates possess an earth sciences or management degree along with wind power industry experience. Project managers typically earn $50,000-$60,000 per year.

Wind Analyst

Wind analysts research weather data relevant to the maintenance and overall performance of the wind farm. If you have a background in earth sciences and physics, a position as wind analyst is a terrific entry-level opportunity, with an average starting salary range of $50,000-$65,000.

Wind Farm Manager

All operations of the wind farm, from the turbines to customer service, employee hiring, training and scheduling, finance and adhering to contractual obligations, are the responsibility of the wind farm manager. While it's unusual to find a candidate with all the desirable educational and experience qualifications, companies will consider applicants with a broader familiarity with wind farm operations, including technical, financial, legal and customer service. Most employers prefer an advanced degree, but those with a bachelor's degree and relevant managerial and industry experience are considered for these exciting jobs. The average salary for a wind farm supervisor or manager is $70,000-$90,000.

WET Career Profile: *Jim Pytel*

Columbia Gorge Community College, located in The Dalles, Oregon, was the first college in the Pacific Northwest to teach wind energy, offering both a two-year associate degree and a one-year certificate program since fall 2007. Around that same time, James Pytel, an electrical engineer with a degree from Clarkston University in New York, had decided to leave the semiconductor field in which he had worked for almost a decade.

"I was bored and looking for something to get me outside," he said. "I had just gotten married and moved to Columbia Gorge. I saw the turbines every where and thought 'Now that's interesting.' So I got a job at the Klondike Wind Farm."

Pytel worked on General Electric turbines as a lead technician, but within six months his wife was pregnant and he needed to find a position that required less travel. CGCC was advertising for a wind energy instructor. Pytel interviewed for the position and began teaching in the fall semester of 2008.

"Our program is unique," said Pytel, "in the fact that it covers electronics, mechanics and hydraulics in the first year. Ours is a jack-of-all-trades program, whereas many other schools may cover one element only in that first year. Ours was one of the first to combine all three together in the first year."

Pytel, who teaches the basic electronics class in the renewable energy program, believes the other major selling point to CGCC's program is that it is a renewable energy technology program, rather than being confined to wind turbines only. The skills students learn in their classes can be taken to a variety of industries, power or otherwise.

"There are plenty of opportunities to use the electronics you learn," he said. "The electronics of a turbine are the same electronics in the car. In the next twenty years, I think, we are going to see the electrification of our existing commuter transportation system. Our students will be able to apply what they are learning in that market also. It's going to be huge, and they will be ready to translate their knowledge to other areas."

In seeking students, Pytel said he wants to see people with strong mathematical competency. "I don't want to have to re-teach them math skills before I can teach them electronics," he said. "The second priority is a person that can translate schematics to reality and reality to schematics. They need to be visually competent. I need them to be able to look at a schematic and then at the real object and see the connection.

"The third thing is to have a good attitude. I tell students that it's always fair, but it's never easy. I want students who can keep their attitudes up, even when the concepts are difficult or they have to spend some time researching to find an answer."

Some of the students who possess these three characteristics are finding success, not only in school, but in landing internships with nearby wind farms.

"This is the most value to the program," Pytel said. "These students are getting their first year of education, then a summer of hands-on experience and then coming back for the second year and graduating. They are going to be very confident wind technicians. If the internship goes well, they will probably get hired upon graduation."

As a final piece of advice, Pytel tells his students to remain flexible and teachable.

"Nothing is stagnant, including renewable," he added. "I think energy independence is the number one thing we should be moving toward, because it is directly related to homeland security. I tell students, 'The skills you develop today will have to be modified for the developments of tomorrow. Always keep educating yourself.'"

Work Schedules

Since wind turbines operate whenever the wind is blowing, technicians must be prepared to work flexible hours and sometimes extended shifts in occasional climate extremes.

Employers

Employers of technical personnel in just about any industry, but especially those dealing in newly evolving technologies like wind energy, will say that learning continues long after you get that certificate or associate degree, whether or not you pursue additional formal education. Your involvement in a "cutting edge" technology means that you will always be encountering new ideas and innovations—new schematics to study, new skills to develop as well as new equipment and instrumentation to master.

The 21st century's renewable energy challenges have already placed renewable energy technologies among the most exciting and important environmental, financial and political initiatives worldwide. Even at a local level, your career in wind energy will keep you at the center of this dynamic revolution changing the ways industrialized nations will power their economies. In order to succeed, you will need to stay current on the latest computer instrumentation technology, meteorological data measurement and analysis equipment, advancements in power collection and distribution and innovation in the manufacture of wind turbines and related equipment.

WET Employer Profile: *Dave Miller*

Asked what he says to young people about choosing a career in wind energy, Dave Miller replied, "I tell them that they have a chance to get in on the ground floor of an industry which is set to explode."

Miller is human resources manager for Wave Wind Energy, a leading provider of small to medium wind energy project planning, development and maintenance based in Sun Prairie, Wisconsin. "I tell them if they're willing to put forth the effort, the opportunities are endless."

A veteran of more than ten years as a human resources professional for firms as varied as a large supermarket chain and an even larger family footwear retailer, Miller says he likes working in the dynamic new wind power industry because he can get to know employees from the time they are starting with the company, through their development, and in some cases, as they leave the firm.

"How exciting to be part of something that is growing so quickly," he said. "It's also nice to be involved in an industry that really is making a difference in the world's reliance on fossil fuels. And I would be remiss not to mention that I was attracted to this role by the senior members of Wave Wind. They are a great company to work for."

Wave Wind Energy was founded by Tim Laughlin and Robert Heinemann, wind energy and construction industry veterans who both saw the potential for wind energy as a way to transition from fossil fuels to renewable energy. Laughlin is a wind energy industry veteran with a background in steel, heavy crane and wind turbine construction. Heinemann has been the owner of a successful television, radio and wireless tower construction business. Together, they anticipated the challenges wind project developers and service providers would face as the demand for wind turbine tower construction experience and equipment began to quickly exceed supply.

"The industry is so new, so early in its evolution, that Wave Wind hasn't started recruiting for field employees from the technology programs yet," Miller said. "For the most part we've been involved in just the construction side of the work. But that will change as we expand the areas of the industry that we are involved in.

Having the education," he quickly added, "either in a certificate, or with an AAS degree, will only help an individual move up in the industry. They will not only have a head start in the technology area, but will have demonstrated the personal discipline and foresight to prepare for a dynamic new career field."

Because of his personal experience with Wave Wind Energy, Miller was reluctant to suggest a range of starting salaries for the industry or a prediction of the demand for wind energy technicians nationwide over the next several years.

"We're seeing high double-digit increases each year in the usage of wind power. With that comes ever increasing numbers of positions. As these numbers become larger and larger, the demand for every type of employee in the industry will only increase. It will take exponentially more people to develop the projects, to build the units, to wire the units, and to ultimately maintain the units."

Miller said the field is wide open, however, adding that, "I feel that within two or three years we will have a huge shortage of trained employees available and that will definitely make it an employee's market." On the other hand, he observes that "the work we do is not for everyone. For the most part, the employees we hire are for travelling positions where the individuals are sometimes away from home for a month at a time. But we value our employees and their families highly and work hard to be sure our people get home as often as possible."

"It's hard to find things not to like." Miller noted, when asked what he likes most and least about his role in the wind energy industry. "The fast pace, the ever-changing environment is exciting. The economic environment [Spring, 2009] is making things a bit more of a challenge right now as financing for new projects has been held up. But once that lets loose—[it] should be sooner rather than later—it's going to be going fast-paced once again."

His general feeling about the future of wind energy technology in the next five to ten years? Miller responded, "It's absolutely the place to be."

The best employers actively encourage their technicians to stay on top of the learning curve. You will be encouraged to join professional associations, read industry journals, newsletters and magazines and to take advantage of Internet resources providing updated information about industry developments across the country and the world. Employers frequently offer their technicians regular in-house training in safety, as well as instrumentation upgrades and incentives for additional formal or continuing education. And there are always opportunities for rapid advancement for great team players who demonstrate an unwavering attention

to detail, to quality assurance, to meeting company objectives and deadlines and to customer service.

As important as hard skills are, employers often look for applicants who demonstrate soft skills that aren't so readily learned in a classroom or lab. Since wind energy technicians regularly work side-by-side with electrical and mechanical engineers, utility company personnel, construction tradesmen and business executives, employers want workers who are excellent communicators, as well as good technicians. As an employee in any kind of industry, anyone you come into contact with is essentially a customer. Your ability to work efficiently with everyone you meet is at least as important as your technical training.

You must be able to provide clear, concise written or verbal reports whenever the occasion arises, in whatever format may be required—as a descriptive written report, a spreadsheet, database report or schematic. You need to be familiar with terminology common to all aspects of your industry:

- Electric motors and generators
- Electricity and electronics
- Instrumentation and measurement
- Construction schematics or wiring diagrams
- Power generation and distribution

Since wind energy technicians may often work in pairs, in somewhat deserted or desolate environs, they must be self-starters, able to prioritize and complete their day-to-day activities with a minimum of supervision. This ability to take the initiative is prized. Effectively and efficiently getting the job done is one of the most important assets in propelling a wind energy career toward success and advancement.

The industries supporting wind energy technology— including those specializing in large and small turbine design and construction, specialized instrumentation electronics, electrical parts manufacturers and providers of tower climbing safety and personal protective equipment—will all be potential employers of qualified wind energy technicians as renewable energy initiatives continue into the next few decades.

Necessary Skill Sets

Wind energy technology is a new and rapidly developing career field that requires a broad range of skills, knowledge and abilities. Much of this professional background can only be learned with time and experience in the field, but a solid foundation is required for a good head start. In addition to the all-important business and interpersonal skills which include business writing and math, the environmental sciences (weather and climate, physics, and chemistry), working with industrial engineers and a variety of people from other backgrounds, leadership qualities such as project management, etc., lay the ground work for starting a career in wind energy technology.

Hard Skills

To prepare for entry level positions in the wind energy industry, applicants will need training and experience in the construction, maintenance and operation of wind turbines. Specifically:

- Troubleshooting and diagnosis of both large and small turbines
- AC/DC theory
- Fluid power (hydraulics and pneumatics)
- Electric motors and generators
- Basics of networking and computer technology
- Computerized control and monitoring systems
- Composites and composite repair
- Data acquisition
- Operational understanding of high tech-low voltage

You'll need to be comfortable working with Supervisory Control and Data Acquisition (SCADA), the industry standard computerized system that controls the wind tower network for public utilities. Wind energy technicians also demonstrate strong backgrounds in safety training and the use of tower climbing and rescue rigging. Additionally, a wind turbine technician must be comfortable with the idea of working at heights of up to 250 feet, in extreme weather conditions and in isolated areas, and must be capable of lifting up to 50 pounds.

Terminology

The technology of wind energy management is a combination of several disciplines; technicians must be familiar with the terms, acronyms and concepts relating to electricity and digital electronics, computer and other instrumentation, earth sciences, hydraulics, welding and aerodynamics (airfoils). It is also necessary to have a good understanding of the means by which wind power facilities interact with other alternative and conventional power generation systems. Finally, qualified technicians are intimately familiar with the names and uses of various types of high tower safety rigging and rescue equipment.

WET Career Profile: *Jeanna Walters*

Jeanna Walters, graduating in 2009 with an AAS in Wind Energy Technology from Cloud County Community College in Concordia, Kansas, is working hard to remove the barriers, both perceived and real, that keep women from seeking previously male-dominated careers. Her second passion is raising environmental awareness and helping people realize their own responsibility when it comes to the sustainability of the planet's resources. She finds the fulfillment of these two goals in her studies as a wind energy technician.

Prior to her CCCC studies, however, Walters had worked for many years in an office environment. In 1984 she started college to pursue a business degree. But marriage and children interrupted her goals, and it wasn't until 2009 that she completed her degree at Kansas State University.

In the interim years, Walters had spent much time as the single mother of three boys heavily involved in the Boy Scouts of America. She became a First Aid and CPR certified instructor and also became certified as a climbing and rappelling instructor. Though she spent her workdays behind a desk, she loved being outdoors and longed for a career that would allow just that. Being a Scout leader had also heightened her environmental awareness.

When she learned of CCCC's Wind Energy Technology program, she realized that here was a career that would bring all of her interests together.

Going back to college to finish a business degree had been somewhat of a challenge, Walters admitted, after so many years of not being in the study habit. Now Walters was faced not only with the discipline of study, but the newness of studying something completely foreign to her.

"I am not a mechanically inclined person," Walters said. "Each semester, there would be one class that I would look at with trepidation. Once it was electronics, and once it was mechanical systems. This last semester it was hydraulics.

"This is different from anything I've ever done. But once I would get into the class and start learning, I would say, 'This isn't so bad. I can do this.' It's been funny to hear my kids ask me if I've done my homework. Sort of a retribution thing for all those years I asked them that, I guess."

Like most career changers, Walter has not used certain skills she learned in high school and during her early college years in a long time. As a result, physics proved to be her most difficult subject. But this woman who looks forward to hanging off a wind turbine nacelle, 300 feet in the air, wasn't about to let a little math scare her away from the career of which she has long dreamed. In fact, when it comes to her expectations for her future career, the sky is the limit.

"Unless I have a fantastic job offer when I graduate," said Walters, "I'm thinking about going on for an environmental engineering degree. I want to learn more about the environmental side of wind energy: land assessment, wildlife displacement and the after-effects of a wind farm. The more I learn about wind energy, the more jobs I see that are out there. Everyday I think I've made up my mind which segment I want to go into, then I learn something new, and I think 'No, maybe this is the way to go.' So right now, I'm thinking about getting an environmental engineering degree at K State."

Besides being within driving distance, Kansas State is also attractive to Walters because it has a chapter of Women in Engineering. Walters has made it a personal mission to raise women's awareness of job opportunities within the wind energy sector, even going so far as to start a Kansas chapter of WoWE, Women of Wind Energy.

WoWE was born in 2005 and is sponsored by Windustry, a nonprofit organization that "promotes progressive renewable energy solutions and empowers communities to develop and own wind energy as an environmentally sustainable asset," according to the organization's Web site found at www.windustry.org.

WoWE is a volunteer-staffed, non-profit organization that has more than 500 members. As of June 2009, there are fifteen chapters throughout the U.S., one in Canada and has seen its first overseas chapter open. WoWE promotes women in wind energy, providing industry information, networking opportunities and job alerts.

Since starting its Kansas chapter, WoWE's members at CCCC have done presentations at local elementary and middle schools to help spread the word about wind energy. Walters wants children to understand that no matter what career they are interested in, it can be related to wind energy, as she and her classmates realized to an even greater degree at a recent wind energy conference held, of course, in the "Windy City" of Chicago.

Across the country where wind farms are established, emergency response teams are being retrained for confined space and high rope rescue. In certain areas, new highway accesses are being built in order to accommodate the larger trucks required to carry the turbine blades to the construction sites.

"Some manufacturing facilities that were closing due to lack of demand have now been refurbished to make wind turbine components," Walters said. "Whatever career you are interested in can be associated with wind. You want to be a nurse? Some wind farms have medical personnel on site. You want to drive a truck?

Truck drivers are being retrained to drive the bigger rigs to carry turbine components."

Walters said that the current demand for technicians is so high that companies are hiring people with little to no training to fill the void. Graduates of accredited schools, like Cloud County Community College are likely to find themselves one step ahead of those with no training. Walters sees an additional benefit from wind energy to colleges, as well. Classes at CCCC fill before the end of the semester and, in order to help fulfill the government's goal of twenty percent renewable energy by 2030, schools like CCCC are finding grant money easier to access.

Even though she can see many benefits to a variety of industries and employees, in the end, one of Walter's greatest desires is to see more women find job, and life, fulfillment within the wind energy sector.

"At my school, out of seventy wind energy students, only three of us are women," Walters said. "And no women have registered for the incoming freshman class next fall. I don't think that's because of the opposition to women being in a male-dominated field. We haven't encountered much opposition from men. There isn't really the intimidation factor today for women going into men's areas like there used to be. Plus this is such a new field that it started after women had already broken into the historically male-dominated workforce.

"I think the problem is that this career is not being marketed to women, so we formed our chapter to get the word out there that this is a career where women can make a name for themselves professionally. It's a career where they can make a place for themselves, not just monetarily, but ethically and morally.

"This is not a government or big business thing. We are all responsible for the sustainability of our planet. Women have a natural fit in that because we are nurturers. This is an opportunity for women to excel. Women as a gender group have not been interested in mechanical fields, but this isn't just electronics and mechanics. There is a whole wide-open field out there. Women need to jump on it."

Computers and Networking

Wind energy technicians regularly work with advanced, dedicated computerized control systems. That means, in addition to the standard computer applications such as word processing, spreadsheet, email and Internet, technicians are fully trained in the uses of programmable logic controllers (PLCs) and industrial automation. Qualified technicians are trained in the basics of computer networking, including terminology, hardware and software components; protocols and topologies that differentiate between various network systems; and they learn to install and troubleshoot network hardware, software and cable. They will demonstrate a working knowledge of network connectivity, configure network protocols and install and configure network client software. Finally, they'll work with PLCs, which are essentially heavy-duty industrial computers used for digital automation of the electromechanical processes that turn wind energy to electric power and which can communicate over a network to related systems, such as another computer running a SCADA system or Web browser.

Troubleshooting

Wind turbines are designed to harness the wind energy that exists in their particular location. Aerodynamic modeling determines the optimum tower height, control systems, number of blades and blade shape. A technician's jobs include monitoring and maintaining the operational efficiency of the turbines' conversion of wind energy to electricity for distribution. More specifically, that means maintaining the major components of most turbines:

- The rotor component includes the blades for converting wind energy to low speed rotational energy
- The generator component, includes the electrical generator, control electronics

- A gearbox for converting incoming low speed rotation to a high speed rotation suitable for generating electricity
- And the structural support components, which include the tower and rotor pointing mechanism

Troubleshooting the equipment requires competence with electronic measurement devices, working knowledge of wind energy theory and practice, training and experience with the diagnostic tools and equipment specific to wind energy power generation. Technicians must be familiar not only with the turbines, but with the equipment that connects the output to the distribution system. Along with problem-solving skills—the ability to look for and pinpoint problems either with their eyes and ears or with diagnostic devices—technicians must be proficient in reading wiring diagrams, schematics, flowcharts, and statistical analyses, in addition to reading and interpreting gauges. They also must be capable of creating maintenance documentation and writing professional reports or summaries, either by hand or by computer. Maintenance and troubleshooting also means that technicians must be thoroughly familiar with hand and power tools, electrical installation and testing equipment, welding and hydraulic equipment, high tower climbing equipment and the proper uses of personal protective equipment (PPE).

Safety
In addition to normal shop safety equipment such as gloves, welding helmet, goggles, etc., wind energy technicians are trained to use climbing equipment designed to protect from slipping and falling during high tower maintenance work. The various types of equipment include rope, cord and webbing, harnesses, rappelling and ascending devices, slings, self-rescue equipment and rope locking or belay devices, along with special shoes, gloves, helmets and other equipment specific to high tower work. A working familiarity with

the Occupational Safety and Health Administration (OSHA), Environmental Health and Safety (EHS), as well as specific guidelines for working on high towers and working over water, are emphasized. The American Wind Energy Association (AWEA) Website lists industry safety guidelines at its Web site: www.awea.org/standards/iec_stds.html. These are international standards, developed by Technical Committee-88 (TC-88) of the International Electrotechnical Commission (IEC), the recognized international body for standards development activities.

WET Career Profile: *Joel Livingston*

Wind technician Joel Livingston works for General Electric, performing troubleshooting and tower maintenance. His job encompasses a variety of duties, including ordering and changing out parts and completing paperwork, proving again those solid written communication skills are important in every job. In addition to acquiring writing skills, he has other advice for young people considering wind energy as a career.

"I tell them it is a challenging and rewarding job, but it is a job that is becoming software advanced," he said. "This means that you not only need an electrical background, but you also need to have some computer knowledge."

Like many wind technicians, Livingston brings a strong mechanical and electrical background to his job. After high school, he spent a year at college before joining the U.S. Air Force, where he spent the next twenty-two years as a B-1 crew chief and maintenance supervisor for thirty-six B-1 aircraft. His extensive electromechanical experience easily transfers over to the wind turbine field, where it serves him well.

However, potential new hires need to understand that no matter what educational or work background they bring to the industry, on-the-job training is a vital part of wind energy careers. GE has its own school, which trains wind technicians along with other specialties.

"I can say the training I received there was very helpful in learning how to troubleshoot and repair wind turbines," he added.

Like some alternative energy professionals, Livingston was attracted to renewable energy by the growth potential in an emerging utility field. "Wind is a growing business that is going to be around for a long time," he said. " I wanted a job that was stable but challenging, without much travel. The push for clean energy ensures that this type of work will be around for a long time, and the advances in technology ensure that I will be exposed to the most advanced training needed to keep the wind turbines functioning."

Working in remote locations and in difficult circumstances presents challenges to any employee, but Livingston is not deterred by these obstacles. His favorite part of the job is being trusted to go out and troubleshoot a malfunctioning turbine, then ordering and changing parts or fixing the wiring in order to get the turbine back online. While Livingston says there is nothing he dislikes about his job, the weather can be uncooperative sometimes, presenting distinct challenges.

Because of the fulfillment he enjoys in his career, Livingston happily recommends this career to anyone who possesses the motivation and determination to succeed in an exciting career that will grow and evolve over the next decade. The demand for renewable energy will continue drive the building and installation of new wind turbine parks, and the current political and economic environment in the United States will work as a catalyst to this growth, providing innumerable opportunities for employment and job satisfaction, he said.

Customer Service and People Skills

As in virtually any profession, one of the most important things employers look for in potential employees is a dedication to customer service and outstanding "people skills." While wind energy technicians may not be as likely to come into contact with actual public utility customers, they are expected to work effectively with fellow technicians and engineers who may be from very different parts of the United States or around the globe. Success in this area is often the result of a strong sense of humor, attention to the expectations of co-workers or supervisors and a willingness to take the extra steps necessary to achieve excellence both personally and on behalf of your employers.

Communication

The most important "soft skill" required for success as technician in any field is an ability to communicate effectively with supervisors and co-workers. Soft skills usually refers to reasonably good writing and speaking skills, but it also includes traits such as personal reliability, analytical thinking, self-reliance and self-motivation, organizational skills, tenacity, common sense, honesty, an ability to set priorities and work as a member of a team, physical fitness, attention to detail and, above all in this sometimes hazardous profession—a determination never to work while under the influence of alcohol or drugs.

Wind energy technicians must be prepared to stay on top of developments in this rapidly evolving field. That means they must join and participate in professional associations and read alternative energy journals and magazines. They must always be looking for opportunities to continue their professional development by attending manufacturers' training schools or workshops, and taking steps to improve their traditional academic and technical skills via online education or in local classes, seminars, or workshops. The most successful technicians in any field are those who consider themselves lifelong learners.

Additional Skills

One of the most important skills for technicians is a conscious effort to take responsibility on the job—for the safety and well-being of his or her co-workers as well as his or herself, to maintain a clean, tidy workspace and to make sure tools and equipment are clean, well maintained and properly stowed away when they're no longer in use. This professional attitude extends to pride in workmanship, solid time management, maintenance of tools and equipment and a constant dedication to maintaining a high level of quality with regard to work documentation and reporting, communication follow-up and excellent work habits like promptness, diligence and respect for fellow employees.

Conclusion

Wind energy technology is one of several renewable energy resources with the collective potential to profoundly change the way human beings power their lives and their economies in the coming decades. By joining the ranks of certified wind energy technicians now, you will be in the vanguard of a profitable, important industry that is already global in scope and unlimited in potential. In the next five to twenty years, experienced wind energy technicians will literally be in a position to write their own tickets, with plenty of exciting, interesting work anywhere in Texas, the United States—or anywhere in the world.

CHAPTER TWO

Wind Energy Technician Education & Certification

By the end of 2007, the wind energy industry in the United States had begun making substantial process on its overall objective of providing as much as twenty percent of the nation's electricity needs by 2030. In order to accomplish that, the industry must triple its output from the current manufacturing and installation base of 5,000 additional megawatts per year. More than 75,000 new turbines will have to be manufactured, installed, connected to the nation's power grids and optimally maintained during the coming decades.

The fact that wind energy is a relatively new technology means that there is a large demand for parts, equipment and labor. Turbine manufacturers face supply shortages of components such as gear boxes and generators. Production of large machined parts like castings and bearings does not yet come close to the high demand, and turbine tower manufacturers are slowed by shortages of rolled steel tower sections and rings. Construction contractors say that construction schedules are regularly slowed by long waits for transformers and the other specialized equipment needed to build the projects. Only a few companies are capable of transporting large wind turbine components.

All these facts result in tremendous opportunities and challenges for new entrants into the industry.

There is also increasing demand for wind energy-aware service industries, such as construction, transportation, legal, financial, safety operations and maintenance. Another result is that the man or woman considering a career in wind energy is presented with myriad career options.

WET Employers Profile: *Sterling and Roland Ramberg*

Sterling and Roland Ramberg of Seattle, Washington, were born to parents who lived the ethics of the old school: ingenuity, hard work and integrity. Their father was employed by Western Gearworks during World War II. When the war ended, the elder Ramberg was able to pick up a surplus gear machine, and he went to work for himself, setting up shop in the garage.

"He served, and we still do, a variety of industries and energy sectors, both hydroelectric and nuclear," said Sterling Ramberg, president of The Gear Works. "We make conveyor gears, high-speed gears, etc. We are the engine room for heavy industry."

Sterling and his brother Roland, who is the CEO and chairman of The Gear Works, started working for their father when they were in high school and have stayed with the family business ever since. They began working with the wind industry ten years ago when they first got inquiries from wind companies in the United States. Since then, the sector has grown until it is twenty percent of the company's total business. "Definitely, it was a great business opportunity," Sterling Ramberg noted, "since we already made high end gearing and always have. Our quality is very high and people know our reputation.

"We do work for Florida Power and Light, enXco and Terra-Gen wind farm owners. Also for Moventus, they are a gearbox OEM (original equipment manufacturer). We recondition their gears. And we do main rotor shaft work for Siemens. The end users are a natural fit for us, but we do have relationships with OEM."

While The Gear Works' 120 employees enjoy the financial benefits of a prosperous business, being involved with a green industry brings them added enjoyment.

"Clean energy is very appealing to our employees and families," Sterling Ramberg said. "We live in Seattle in the Northwest. We appreciate being able to eat a fish that is clean and won't make us sick. It's very exciting to be a part of the clean energy movement."

Additionally, working for a family-owned business fosters employee loyalty. "Many of our employees have been here for a long time," Sterling Ramberg noted. "One man just retired after thirty-nine years. We take a long-term view; it's in our blood. We grew up in the gear business and the employees are part of the family. When you're part of the family, you stick around longer than when you're a number somewhere. It makes sense to keep our talent, which we invest a lot in, rather than having a situation where I have to chop heads to make our numbers for stockholders."

And being a family-owned business makes The Gear Works unusual in the market, he added. "You look at the wind industry jobs directly and there's 'Oh I go up in the tower and work on the farm.' We're not on the farm; we focus our energies on making good, quality gears and making improvements to gearbox assemblies. That is vital to keep these fleets running.

"Typically, on a wind farm the OEMs are pushing larger and larger models because it economically pencils out better to make the larger units. Gearboxes that weigh twenty tons are becoming the small ones. So the existing fleets of the smaller units are saying 'Who's going to take care of me?' That's where we're especially going to be needed. There are only a few shops like us that custom make everything. We don't make only one type. We are flexible and adaptable. Not only can we do the disassembly and analysis, we make the part. We use the same equipment that the OEMS in Europe are using. Quality is not compromised. We are unique in that we have the assembly capacity and the gear-making capacity all in one."

Although a profitable business, it is not one without its risks. Ramberg describes the business as capital intensive and quoted one of his father's sayings, " 'There's a limit to how much you can make, but no limit to how much you can lose in this business.' We take big business risks all the time to do this. You must bring in new technology and advance your machining tools or you will perish in this business."

Yet with the Barack Obama administration's goal of having twenty percent of the nation's energy coming from renewable sources by 2030, combined with the fact that companies like Siemens are looking to purchase all American-made components, it is likely that The Gear Works will continue to see growth in the wind energy sector of their business.

Ramberg advises people interested in applying for a position anywhere in the wind industry: "You have to have a reasonable background in math because we are making and assembling parts to very exact standards. So you must have mathematical and mechanical aptitude. There are local trade schools we work with here to identify apprentices. We have about six [apprentices] right now, which is quite a few for a gearbox company of our size. Real training happens here. Even gear engineers learn engineering in the real world, not in college."

Ramberg also views his business as something of a hospital. "The wind turbine's nucleus is the gearbox. Engineers are still trying to figure out what kind of loads they see in the harsh environment they are in. They need taking care of. That's what we do. And not only are we a hospital that fixes things, we improve them."

Educational Requirements

As of early 2009, fewer than a dozen technical colleges or universities offered two-year AAS degrees in Wind Energy Technology, the majority of them in the Midwestern tier of states from Minnesota to Texas. Although most are still in the planning phase, several universities will soon be offering four-year and post-graduate degrees in WET or management, most often as a specialization track subsequent to major studies in electrical or industrial engineering.

Many schools and businesses currently offer introductory short courses or WET orientation events on an irregular basis. Typical of these is the two-day "Wind Energy Course

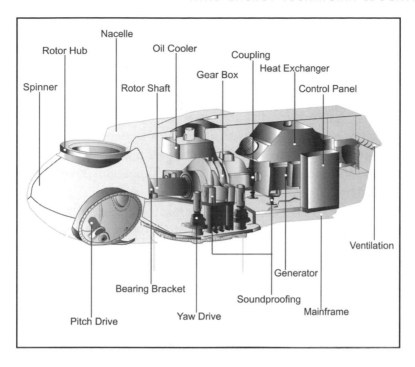

for Technicians" at the University of California-Davis. This particular program features discussions and limited field orientation programs such as an introduction and historical survey of the various types of structures that have been developed to harness the wind, from medieval European windmills and water pumps to modern utility-scale turbines. Other subjects include an examination of wind characteristics over various types of terrain; the aerodynamics of wind turbines; a study of the structures and loads on wind turbine components, from the airfoils and rotor to the tower foundation; the engineering rationale behind turbine operation and maintenance procedures, and performance analysis. Also part of the workshop are brief surveys of wind power electrical systems and grid integration, as well as the environmental issues involved in siting large wind farms—such as noise, visual impact, agricultural use complications, aviation lighting requirements, electromagnetic radiation and lightning protection. (Information about the UC-Davis course is available at cwec.ucdavis.edu/training/)

Many of these short programs focus on regional concerns or specific industrial applications of WET. The American Wind Energy Association (AWEA) maintains events calendars at www.awea.org/events/ and www.awea.org/events/other industry_events.html. The United States Department of Energy maintains an extensive list of seminars, short courses, and public and private academic certificate and degree programs in WET at www.windpoweringamerica.gov/schools_training.asp.

WET Employer Profile: *Ryan Dawes*

Headquartered in India, Suzlon Energy Limited has a presence in twenty-one countries around the world with manufacturing facilities in India, China and the United States. It is the fifth largest manufacturer of wind turbines and, additionally, Suzlon operates wind farm sites throughout the world, including in the United States.

Ryan Dawes is the manager of one such farm, operated by Suzlon's American subsidiary, Suzlon Wind Energy Corporation (SWECO), in Oregon. As a manager of an emerging technology, Dawes' day begins early and ends late, with an hour commute each way added on to his day. He said the commute gives him time to go over the seven a.m. meeting he conducts when he arrives at the job site.

After his meeting, he spends time with individual employees who need his input. The rest of his day is occupied in answering emails, maintaining and ordering parts inventory and keeping his techs well supplied to optimize the day-to-day operations of a wind farm housing forty-seven of the Suzlon S88-2.1 MW turbines.

There was a time in Dawes' life that, although he tinkered with the idea of renewable energy, he relegated the notion to the hobby arena. A former U.S. Air Force avionics technician who later moved to civilian work with first Fed-Ex, then Gatorade as a distribution manager, Dawes has always been a "hands-on" employee. His interest in wind energy began in his own garage, where he built his own solar power station after reading the do-it-yourself magazine, Home Power.

"Back then," he said, "I never thought there would be a renewable industry that would actually provide a living wage; I thought it was just a hobby. But then it started going, and I saw so much on TV about it, with T. Boone Pickens' commercials, etc. I was looking for a new job at the time, going through a hiring agency, and they introduced me to Suzlon. I was very happy once I realized I could have a career in renewables."

Today, Dawes manages seven technicians, and one lead tech. "The lead tech is my go-between," he said. "Sometimes he's in the field with his men, and sometimes he's in the office with me, helping me with parts, tools or making assignments. I'd like to be in the field a little more, but as a manager, my job is mostly in the office."

And although he enjoys the interaction that he gets with his men during the early morning meeting where laughter and camaraderie is prevalent, sometimes his day brings the not-so-pleasant responsibilities of management.

"Because this is such a new industry," Dawes said, "it's very much an overtime environment. One of my biggest challenges is keeping the employees satisfied because they have to work so many hours. It's never pleasant having to tell someone they have to work on the weekend."

Yet the potential earning power achieved through overtime is exactly what brings many technicians into the field where starting salaries nationally can range from $15 to $20 per hour, according to a March 1, 2009, *Los Angeles Times* article. And Dawes is certainly not one to ask his men to go where he isn't willing to go. Most of his days range from ten-to-sixteen hours, depending on the circumstances of the day, and that's not counting his two-hour commute.

The long hours don't dampen his excitement over his career, however. "Suzlon is a great company to get in with; it has a lot of potential," he said. "Any company starting out in the United States has such growth opportunity if you get in on the ground floor. The amount of power that a single turbine can put out has increased from a few 100 kilowatts per hour up to two-to-five megawatts. They [the turbines] are becoming more competitive with other forms of energy. I think I have a great career ahead of me."

It's a career he recommends to anyone who has a natural aptitude for electromechanical skills. "Get into a good school that teaches turbines and definitely take up an intern position if it's available," he recommended. "Get your hands dirty; see if you like it. Talk to professionals. Get a taste for it. Don't just read about it in books."

Although Suzlon currently does not require college degrees for its technicians, it is a direction that this company, like most other wind energy companies, is heading. "They [Suzlon] are starting to look at requiring an associate degree, and at a minimum, some kind of technical training, so it's important to go to school. A strong electrical background is important. Of course, mechanical skills are, too, but electrical skills are primary. You must have good trouble-shooting skills and apply what you know to the real world. You have to figure out why this turbine is doing what it's doing and get it back in operation as soon as possible."

Dawes also recommends looking at *Home Power* magazine if a person is interested. "Try out some different do it yourself kits. Teach yourself and your kids a little bit about it, and see if you're interested, then contact our HR [human resources] division or check out our Web site for job openings."

Suzlon can be found online at www.suzlon.com.

The Two-Year Associate of Applied Science Degree

The coursework leading to an AAS degree in WET or management typically consists of introductions to the principles of DC and AC circuits, computer applications and networking, wind turbine construction materials and electromechanical equipment operation and maintenance. Graduates are trained in fundamentals of digital instrumentation, industrial automation and programmable logic controllers and their applications to wind energy generation and distribution. Basics of reading manuals, blueprints and schematic diagrams, fluid power (hydraulics and pneumatics), wind turbine troubleshooting and repair and, of course, the business of wind energy are important facets of the training. The stronger programs also offer training in career skills such as composition, interpersonal communication, applied sciences and math.

An important and unique aspect of training for technicians operating and maintaining wind turbines and related equipment—one which is integrated into virtually every phase of a technician's training—is job safety. Every degree-granting program should include not only general safety knowledge and skills, including Occupational Safety and Health Administration (OSHA) standards and personal protective equipment (PPE), but the more practical applications of harnesses and helmets, lanyards and connections, uses and construction of various types of rope, safety systems and knot-tying. Graduates must be thoroughly grounded in climbing safety, work positioning and efficiency, safe anchor systems, as well as techniques for ascending and descending towers and structures. Adequate training should be provided in rescue techniques, first aid and teamwork required for complex rescues.

Finally, organizational safety is essential to superior job preparation for wind energy technicians and their supervisors:

- Hazard analysis
- Development of safety plans and controls
- Identifying and understanding safety resources
- Clear and concise communication channels
- Budgeting and allocation of risk management and safety resources
- Ongoing training

The key to finding a successful WET education program is to speak with faculty members, graduates of the program and employers of the program's graduates.

The typical two-year AAS academic program in WET consists of two full-time years of study, plus one or two summers of field training, for a total of forty-eight to sixty semester hours. Ideally, the courses should be heavily lab-oriented, providing students with ample hands-on experience and training, preferably led by an instructor who has spent time in the industry. Students in this type of program very often have jobs waiting for them upon graduation.

Not many years ago, the associate of applied science degree was considered a "terminal" degree. Most graduates did not intend to continue on to a four-year program, in part because transferring credits was difficult. Today, many traditional colleges are offering "inverted" degrees in which an individual can transfer a block of technical course credits (typically thirty-six hours) and common "core" academic courses (another thirty to forty-eight semester hours) and then take the upper-level courses online, sometimes finishing the degree in as little as two years. These degrees may have names such as bachelor of applied technology, bachelor of applied science and technology, or bachelor of applied management. They often capitalize on the technical background earned in completing the two-year degree and add leadership and management skills. These inverted degrees offer the fastest and least expensive track to a valuable four-year degree.

In addition to attracting individuals who are not interested in, or cannot afford, a four-year degree, formal technical training in WET is becoming a necessity for people who are already experienced professionals in the fields of industrial engineering, electrical and mechanical engineering, and public utilities. The emergence of the high-profile "green energy" resources phenomenon is leading individuals already working in the private or public energy industries to enroll in one- and two-year programs to upgrade their skills and knowledge about renewable energy technologies such as solar, hydropower, geothermal as well as wind power. As a result, WET is one of those rare "cutting-edge" technologies that offer valuable training to entry-level employees as well as to returning engineering and business professionals who need to beef up their professional qualifications.

WET Career Profile: *Lucas Chavey*

Lucas Chavey, like many wind professionals, changed careers to come into the field. He holds a bachelor of science degree in physics, which he earned in Kansas. After college, he moved to San Antonio, Texas, to work for a while. Eventually he returned to Kansas at about the same time the wind farm buzz was gearing up in his area.

Chavey started reading the articles in the local papers and some of the industry journals, and he was increasingly intrigued. Also at this time, Cloud County Community College in Concordia, Kansas, was offering its first wind energy classes, so Chavey signed up for the program. While a student in the wind energy program, he worked for the Balance of Plant constructor Michels Corporation in an internship position arranged by CCCC where he worked with the project management team.

Before long, Cloud County Community College was searching for a new wind energy instructor, someone who had both education and experience. Chavey applied for the job and began teaching at CCCC in 2008.

The only college in Kansas that is approved to offer degrees in wind energy technology, CCCC offers a two-year associate degree in applied science of wind energy technology. For students who don't want to take two years of classes before getting into the workforce, the college also offers a one-year certificate program. Students who take the two-year program incur more general education classes, of course, but they also take more upper-level classes, such as transformer theory and Supervisory Control and Data Acquisition (SCADA)—the computer control system used in all wind turbine systems applications.

The popularity of the CCCC program grows each semester. Since the time that Chavey took his first classes at Cloud County, the program has doubled and, at times, tripled in size each semester.

When interviewing students for the program, Chavey said he

looks for people who are in good physical condition for climbing 200-300 feet in the air. "I tell them in order to fix the problem on the turbine, you have to be able to get to the job site, and the job site is usually in the air," he said. "I'm also looking for a student who has the ability to adhere to safety rules and guidelines. These are the two definite essentials that no employer will consider you for a job without. Safety is extremely important."

Additionally, Chavey looks for students with technical aptitude, mechanical knowledge and an understanding of electrical concepts, such as circuitry and three-phase power generation. Problem solving is another important trait that a potential wind turbine technician needs to possess. "I tell students they have to be a problem solver because you are troubleshooting a turbine when it's not working, or you are optimizing performance. Best case scenario, you are doing simple maintenance. But it's just you and one other person out there, and whatever needs to be taken care of must be done while you're there.

"Wind turbines epitomize all types of systems. Hydraulics, mechanical, electrical, computer; you have to be able to fix the problem while you're there, and you have to know all the systems of a turbine," he added.

Because today's turbines and wind farms are controlled with elaborate PLC (programmable logic controller) systems, much of the troubleshooting is done with a laptop. Students must be computer and software savvy, too.

"There's a lot to learn, and all turbines aren't exactly alike," he said. "I tell students that they must be able to take what they've learned in the classroom and apply it in advanced, real-life situations. The industry is so new that there is not a lot of standardization. So the only way to be successful in this career is to have a strong understanding of the fundamentals, and that's what we teach in our classrooms."

In addition to giving students a strong grounding in the fundamentals of turbine maintenance and repair, CCCC prides itself on having a live lab situation where students practice their skills. A Kansas utility company donated a nonfunctioning turbine to the school, with the understanding that the students would make the turbine operational. This gives students access on a daily basis to the very structure they will be working on in their careers.

Chavey said, "On a nice day, we can go out and practice climbing, or rescuing, or look at the systems inside the nacelle. Because the hazards like high-voltage are removed, with this being a nonfunctional tower, we don't have to worry about someone forgetting lockout/tagout procedures and a piece of equipment suddenly coming on and hurting someone."

The college isn't stopping with the one turbine however. "We'll also be creating a power purchase agreement with utility company in Kansas," Chavey said, "and we will have a two megawatt turbine putting energy back in the grid. And we'll have a smaller turbine on campus to generate power for the school."

The timing for this planned growth and availability of equipment could not be better, as Siemens, a leader in the renewable energy sector, is planning a nacelle manufacturing facility in nearby Hutchinson, Kansas. The plant will begin shipping its first nacelles in 2010, and CCCC expects that the company will employ many of their graduates and interns. Siemens forecasts needing 160 employees at start-up, growing to 400 employees at full capacity.

The industry growth experienced in Kansas is predictive of the U.S. market as a whole. "I think it [the industry] is set to explode," Chavey said. "The United States was the number one market in the world last year [2008] for wind energy installation. With our current economy, that has slowed some, but forecasters expect it to ramp up again by the end of this year [2009] or first of next. The U.S. government's goal is to have twenty percent of our energy coming from renewable sources by the year 2030. Wind will be a big part of that. It's predicted that by 2030, there will about 30,000 wind energy technicians needed. That's 1,500 per year who need to be educated."

And that number refers to technicians only. Wind energy spins off a host of industry related jobs, including manufacturing, construction, transportation and safety. When all aspects are considered, the Bureau of Labor forecasts approximately five million new jobs will be created by the industry over the next twenty years.

Entry-level salaries will fluctuate, of course, based on the type of job an employee has and for whom he or she works. But Chavey believes his technician students can expect a starting range of $17 to $20 per hour, based on two scenarios.

"Salaries are driven by two things," he added. "What part of the country you work in and what type of technician you are. Some technicians are dedicated to a particular wind farm. They learn that company's turbines and specialize in their systems only. Other technicians are special project techs. They travel nearly 100 percent of the time. They can be generalists, or they might specialize in a certain system, like the computers, or gearboxes or maybe electrical. Typically the ones who travel all the time make a little more."

Students often ask Chavey if they will earn a higher entry-level salary if they take the degree program, as opposed to the one-year certificate program, and the answer is probably not.

"I tell them the difference comes in momentum," Chavey said. "Students who have the degree are much more likely to move up quickly to management or lead positions in their careers. Not being willing to put that extra year into your education (now) can cost you hugely down the road in your career options. It's always worth it to advance your education."

The Four-Year Bachelor's Degree

Individuals aspiring to supervisory and management careers usually find that a four-year degree is necessary. The degree major is a matter of preference, but business administration, computer networking and information systems, industrial manufacturing and/or engineering technologies are all good choices. The choice of degree will finally depend on the technician's current degree, future plans, length of time or finances he or she wishes to spend to earn the degree, as well as the availability of online or in-class courses.

Anyone interested in a four-year degree should verify that the granting institution is accredited by one of the six regional accrediting associations in the United States.

Accreditation assures that your credits will most likely be transferrable to other colleges—such as a graduate school—and be accepted by the best employers.

These accreditation associations are the New England Association of Schools and Colleges (NEASC), the North Central Association of Colleges and Schools (NCA), Middle States Commission on Higher Education (MSCHE), Southern Association of Colleges and Schools (SACS), Western Association of Schools and Colleges (WASC), and Northwest Commission on Colleges and Universities (NWCCU). Most colleges clearly state their accreditation by one of these six regional associations on their Web sites. Claims to be licensed only by a certain state or other claims to accreditation should be considered red flags and schools making these claims, or not stating any kind of accreditation at all, should be avoided.

WET Employer Profile: *Joe Brenner*

In 1995, Nordex produced the first megawatt wind turbine system in the world. Since then, this pioneer in turbine technology has remained a leader in the industry, introducing the first 2.5 MW turbine in 2000. Nordex's Web site (www.nordex-online.com) reports that today there are more than 3,857 Nordex wind turbines with a total rated output of more than 5,000 megawatts rotating in 34 countries of the world. The company has offices and subsidiaries in eighteen countries.

Currently, Nordex USA Inc. is planning a plant in Jonesboro, Ark., that will manufacture every part of a wind turbine except for the tower, which includes the blades and nacelles. Joe Brenner, vice president of production, will oversee the plant.

Brenner, a mechanical engineer, began his career in his family's contract manufacturing business. In 2005, however, as part of the lingering effects of the 9/11 terrorist attacks on America, the manufacturing industry was experiencing a slowdown, so Brenner made the decision for a career change.

"Just as manufacturing was slowing," he said, "wind energy technologies were emerging. It was perfect timing."

Brenner's first job was with a start-up Spanish wind energy company opening its operations in the United States. "I was the second employee hired on the operations side," he said. "So I was able to start from the ground up, being involved in every area from operations to purchasing to manufacturing. I learned a large portion of the wind energy business from those three perspectives."

By 2008, his experience combined with his mechanical engineering degree landed him the VP spot with Nordex. In his position, he is heavily involved with the planning stages of the new plant opening soon in Jonesboro.

"We're involved in the tooling that needs to be prepared, the plant equipment, planning budgets and all the strategies for a sophisticated manufacturing assembly line process," Brenner said of his team. "We're presenting these budgets, strategies and concepts to the board, and we're participating in setting up the critical components. There are twenty-two cranes that must be sourced, which is a big endeavor. Then there's the IT [information technology] infrastructure, telecommunications and security card access. We are also studying the drawings from the design firm to make sure they meet specifications. We're meeting new challenges everyday."

In order to meet those challenges, Brenner works closely with the setup team in Germany, the corporate offices in the United States, the design-and-build team, and the senior project construction team in Jonesboro. Brenner, a true problem-solver like most engineers, meets his challenges with relish.

"I like building a team, mentoring and, in this case, starting on the ground floor, so we can develop our own culture. We won't have to fix anything, as you do when you come into an existing culture, but can build it from the ground up. I also like being involved in an industry that is emerging. Many industries have left the United States, but this one is growing here. Foreign companies are setting up here, creating jobs for Americans. When things slowed down after 9/11, many manufacturers closed. Now we're seeing a resurgence because of wind. There is so much excitement from a manufacturing professional's point of view."

Brenner believes that one of the most important components of anyone's career is to be able to trust that their career is long term. "What we see here (with wind energy) is something that will be sustainable; something that will be here for a longer period of time than most industries. The industry is unique because of the size of the turbines. It's very expensive to import them, so it makes more sense to manufacture and purchase products close to where you're assembling or installing turbines."

When the Jonesboro plant is completed and begins operation, the company will be assembling wind turbine nacelles using primarily parts from Germany, but with a goal of using all American-built parts within four to five years.

"Of course there will be challenges, but it is very satisfying overcoming obstacles. And it's a good problem to overcome. We need to turn that into an advantage for the future workforce in the United States. Take blade manufacturing, for example. It requires a specialized skill set, and the pool of skilled workers is extremely limited. You have to train employees from the ground up because it's unlikely they will come to you with prior experience from another industry. Nacelle building is heavy industrial work, so quite often there isn't a group of employees who can move from one industry to this one without training."

Already on top of this problem, Nordex is making advanced preparations to deal with the need to train a workforce. They have planned for a training academy next to the plant, where incoming employees will be submerged in the company's culture and necessary skills. Brenner added that Nordex's goal is to train employees and have their skills "ramped up at the rate of demand."

He said, "We must have a quality workforce at the same time that we're responding to demand. We have to make an investment in people. It's not just a $100 million facility. The most important resource we can put time and money into is the workforce."

With that in mind, Brenner advises potential employees to come into the interview with a great attitude.

"One of the key things I've seen," he said, "is when you interview an employee who is as excited as you are when you talk about wind energy. You can't help but get more excited, too. The employees must have the same interest, desire and passion that we want on our team. Of course, they need mechanical and technical skill sets, and thirdly they have to be reliable. But that intangible is a high interest level in an emerging industry. Put those three things together, and that's the perfect employee."

The enthusiasm Brenner wants to see in his employees is hard to miss in his own conversation. "Everyone is in an anticipation mode," he said. "This [industry] is not a novelty; it's really going to happen. There are just too many factors that are indicating the industry will be strong. We have the support of the [presidential] administration, a constantly increasing need for energy and climbing oil prices. The wind industry now is becoming much more efficient in terms of output, and it continues to grow.

"There's a high level of interest from the whole United States; anyone you talk to in any field sees this as a very positive thing. It crosses the boundary of both political arenas, and there's a consensus among everyone that this is here to stay. So we at Nordex must be able to execute our plan and get ready for the big surge that definitely will happen."

Tuition & Fees

It is not unusual for industry employers to offer tuition assistance or reimbursement for their technicians to help offset the costs of obtaining a four-year degree. With many colleges now offering courses online, working as a wind energy technician full time and pursuing a four-year degree part-time with an employer's assistance would certainly ease the financial strain of additional formal education. If the employer offers a substantial tuition benefit, there are a large number of possibilities for online programs to enable wind energy technicians to earn a higher degree.

For those whose employers do not provide tuition assistance, there are still many economical options for earning a four-year degree from good universities. Many states have public universities with low tuition rates for in-state residents. An excellent web site for investigating accredited online programs is www.geteducated.com/.

Continuing Education

WET is a rapidly evolving technology, with new turbine designs, computerized control systems and power grid distribution developments being reported globally on an almost constant basis. Wind energy technicians should not only be aware of the new projects that are being planned or constructed in the region, but must keep up with the advancements in wind energy and related technologies to remain valuable as an employee. The best technicians read trade magazines and journals such as AWEA's *Wind Energy Weekly* (monthly, by subscription: www.aweastore.com/), *Windletter* (electronic publication, by subscription: www.awea.org/membercenter/), *Wind Power Monthly* (www.windpower-monthly.com), and *Renewable Energy News* (energy.sourceguides.com/news.shtml). These and other industry publications help technicians to stay informed about the latest technical and business developments in the wind power industry.

At this time, the primary national wind energy professional association is the American Wind Energy Association (AWEA), a trade and advocacy organization based in Washington, D.C. AWEA represents the U.S. wind energy industry and individuals who support clean energy legislation and regulation, marketing and public information and international communication and cooperation. The Association's main Web site is www.awea.org/.

Many wind energy technicians will have the opportunity to attend manufacturer training on specific equipment through their employers, with the cost of this training often included in the employer's purchase of the equipment. However, most technicians may spend up to six months on the job before being considered for attendance at manufacturers' training schools.

Finally, as an employee in any field, it's a great idea to develop the attitude that learning is a life-long responsibility. No one, no matter how educated or accomplished or highly trained, can say he or she already knows all there is to know about any job. There is always more to learn, technical and "soft" skills to polish, as well as new developments to learn about and apply to daily responsibilities. Most communities offer seminars, short courses, community college or technical college classes that will enable most technicians to constantly upgrade their education and training. You may find it worthwhile to take formal classes to improve your skills at writing business reports or your understanding of statistics or your facility with computer applications. Continuing education not only keeps your personal marketability up to date, but it's a great way to meet interesting, influential people in your profession and your community.

CHAPTER THREE

Additional Wind Turbine Equipment Technician Information & Resources

WET Higher Education Programs in the United States

Arizona

Arizona State University
Tempe, AZ
www.asu.edu
(480) 965-9011

Northern Arizona University
Flagstaff, AZ
Thomas Acker
home.nau.edu
(928) 523-9011

California

Cerro Coso Community College
Ridgecrest, CA
Sheryl Plett
cerrocoso.edu
(760) 384-6100

Shasta College
Redding, CA
Sharin Truett
www.shastacollege.edu
(530) 242-7500

Solano Community College
Fairfield, CA
www.solano.edu
(707-864-7000

University of Calfornia, Davis Extension
Davis, CA
Dr. C.P. van Dam
training@cwec.ucdavis.edu
cwec.ucdavis.edu/training
(530) 752-7741

Colorado

Colorado State University
Fort Collins, CO
Michael Kostrzewa
michael.kostrzewa@colostate.edu
www.engr.colostate.edu/me/pages/energy_env.html
(970) 491-7709

University of Colorado
Boulder, CO
Martin Dunn
www.colorado.edu
(303) 492-1411

Delaware

University of DE
Newark, Delaware
Willett Kempton
www.udel.edu
(302) 831-2792

Florida

Florida State University
Tallahassee, FL
www.fsu.edu
(850) 644-2525

Illinois

Illinois State University
Normal, IL
David Kennell
www.ilstu.edu
(309) 438-2111

Idaho

Boise State University
Boise, ID
Todd Haynes
toddhaynes@boisestate.edu
coen.boisestate.edu/WindEnergy/WfS/index.asp
(208) 426-4053

Iowa

Des Moines Area Community College
Ankeny, IA
www.dmacc.edu
(515) 964-6200

Iowa Lakes Community College
Estherville, IA
Angie DeJong
adejong@iowalakes.edu
www.iowalakes.edu/programs_study/industrial/
wind_energy_turbine/index.htm
(712) 362-8374

North Iowa Area Community College
Mason City, IA
www.niacc.edu
(641) 423-1264

University of Iowa
Iowa City, IA
Tim Rocheleau
trochele@engineering.uiowa.edu
www.mie.engineering.uiowa.edu/IEProgram/
WindPowerManagement.php
(319) 335-5668

Kansas

Cloud County Community College
Concordia, KS
Bruce Graham
bgraham@cloud.edu
www.cloud.edu
(800) 729-5101

Highland Community College
Highland, KS
www.highlandcc.edu
(785) 442-6000

Kansas State University
Manhattan, KS
Ruth Douglas Miller
rdmiller@ksu.edu
www.ece.ksu.edu/psg/wac
(785) 532-4596

Maine

Northern Maine Community College
Presque Isle, ME
Wayne Kilcollins
www.nmcc.edu
(207) 768-2700

Massachusetts

University of Massachusetts
Amherst, MA
James Manwell
Jon McGowan
www.umass.edu/umhome
(413) 545-0111
(413) 577-1249

Michigan

Delta College
University Center, MI
info@delta.edu
www.delta.edu/home.aspx
(989) 686-9000

Kalamazoo Valley Community College
Kalamazoo, MI
Debbie Dawson
www.kvcc.edu
(269) 488-4400

Lansing Community College
Lansing, MI
(517) 483-1876

Lawrence Technological University
Southfield, MI
Dean Devdas Shetty
www.ltu.edu
(284) 204-4000

Minnesota

Minnesota West Community and Technical College
Canby, MN
Gary Olson
gary.olson@mnwest.edu
Associate of Applied Science, WET
www.mnwest.edu/programs/program-type/aas/
wind-energy-technology
(507) 223-7252
Associate of Applied Science, Wind Energy
Mechanics
www.mnwest.edu/programs/program-type/
diploma/wind-energy-mechanic

Katie Delaney
katie.delaney@mnwest.edu
Certificate, Windsmith
www.mnwest.edu/programs/program-type/
certificate/windsmith
1-800-658-2535

Mesabi Range Community and Technical College
Eveleth, MN
Dan Janisch
d.janisch@mr.mnscu.edu
www.mr.mnscu.edu
(218) 744-7518

Missouri

Crowder College
Neosho, MO
www.crowder.edu
(417) 451-3223

Pinnacle Career Institute
Kansas City, MO
Edwin French
www.pcitraining.com
(806) 331-5700

Montana

Montana State University
Billings, MT
www.msubillings.edu/cot/programs/prog
autotech.htm
(406) 657-2011

Montana Wind Application Center
Montana State University
Bozeman, MT
Mike Costanti
mconstanti@westerncommunityenergy.com
www.coe.montana.edu/wind/

New Mexico

Mesalands Community College
Tucumcari, NM
John E. Hail, Jr.
johnh@mesalands.edu
www.mesalands.edu/wind
(575) 461-4413 ext. 156

New York

Clinton Community College
Plattsburgh, NY
Janice Padula
www.clinton.edu
(518) 562-4200

Cornell University
Ithaca, NY
as.cornell.edu/index.cfm
(607) 254-4636

North Carolina

Appalachian State University
Boone, NC
www.appstate.edu/
Dennis Scannlin
(828) 262-2000

North Dakota

Lake Region State College
Devils Lake, ND
www.lrsc.edu
(701) 662-1600

Ohio

Case Western Reserve University
Cleveland, OH
David H. Matthiesen
www.case.edu
(216) 368-2000

Oklahoma

Oklahoma City Community College
Oklahoma City, OK
Rhonda Cantrell
rcantrell@occc.edu
www.occc.edu/corporatelearning
WindEnergyCertificate.html
(405) 682-7853

Oklahoma State University
Oklahoma City, OK
Jerry Nielsen
njerry@osuokc.edu
www.osuokc.edu/engineering/default.aspx
(405) 945-3222

Oregon

Columbia Gorge Community College
The Dalles, OR
RET Advisor
www.cgcc.cc.or.us
(541) 506-6011
(541) 308-8211

Lane Community College
Eugene, OR
Debra Ganser
www.lanecc.edu/collegecatalog/documents/
0809catalog.pdf
(541) 463-5034

Pennsylvania

Saint Francis University
Loretto, PA
www.francis.edu
(814) 472-3000

South Dakota

Mitchell Technical Institute
Mitchell, SD
www.mitchelltech.edu
(800) MTI-1969

South Dakota State University
Brookings, SD
Michael Twedt
michael.twedt@sdstate.edu
South Dakota Wind Application Center
wac.sdwind.org
(605) 688-4303

Tennessee

University of Tennessee
Knoxville, TN
www.utk.edu
(865) 974-1000

Texas

Clarendon College
Childress, TX
Donny Cagle
www.clarendoncollege.edu
(940) 937-2001

Amarillo College
Amarillo, TX
Jack Stanley
jbstanley@actx.edu
www.actx.edu
(806) 371-5091

Howard College
Big Spring, TX
Tom Land
www.howardcollege.edu
(453) 264-5000

Midland College
Midland, TX
Tracy Gandy
www.midland.edu
(432) 685-4500

South Plains College
Lubbock, TX
Raymond Elizonzo
www.southplainscollege.edu/home
(806) 894-9611

Texas State Technical College – Harlingen
Harlingen, TX
Enrique Carrillo
www.harlingen.tstc.edu
(956) 364-4733

Texas State Technical College – West Texas
Texas Wind Energy Institute
Abilene, TX
Georgan Hausenfluke
georgan.hausenfluke@abilene.tstc.edu
(325) 734-3684

Texas Tech University
Lubbock, TX
Dr. Sukanta Basu
sukanta.basu@ttu.edu
www.wind.ttu.edu/index.php
(888) 946-3287

Western Texas College
Snyder, TX
www.wtc.edu
(325) 573-8511

West Texas A&M University
Canyon, TX
www.wtamu.edu
(806) 651-0000

West Virginia

Eastern West Virginia Community and Technical College
Moorefield, WV
Ward Malcolm
wmalcolm@eastern.wvnet.edu
www.eastern.wvnet.edu
(877) 982-2322
(304) 434-8000

Wisconsin

Fond du Lac Tribal and Community College
Fond du Loc, WI
admissions@fdltcc.edu
www.fdltcc.edu/academics/departments/certificate/
clean-energy-technology.shtml
(218) 879-0805

Lakeshore Technical College
Cleveland, WI
Douglass Lindsey
gotoltc.edu/programs/windEnergy.php
(920) 693-1265

University of Wisconsin
Madison, WI
www.wisc.edu
(608) 263-2400

Wyoming

Laramie Community College
Cheyenne, WY
Mike Schmidt
mschmidt@lcc.wy.edu
www.lccc.wy.edu/Index.aspx?page=1092
(307) 432-1639

WET Two- and Four-Year Degree Plans

WET Degree Programs: Texas State Technical College West Texas (Sweetwater, Texas)

	Texas State Technical College	
Students Starting Fall 2010	Wind Energy Technology Two-Year Degree	Total Credits: 61
First Semester		
LEAD	Corporate & Community Development with Critical Thinking	2
WIND	Introduction to Wind Energy	3

CETT	Electricity Principles	4
MATH	College Algebra	3
ENGL	Composition I	3
	Semester Totals	**15**
Second Semester		
WIND	Wind Turbine Materials and Electromechanical Equipment	3
CETT	DC-AC Circuits	4
ELMT	Basic Fluid Power (Hydraulics and Pneumatics)	3
ITNW	Network +	3
POFI	Computer Applications I	3
	Semester Totals	**16**
Third Semester		
WIND	Wind Business	3
WIND	Wind Power Delivery System	4
INMT	Industrial Automation	4
	Humanities/Fine Arts Component	3
	Semester Totals	**14**
Fourth Semester		
POFT	Job Search Skills	1
WIND	Turbine Troubleshooting and Repair	4
ELMT	Programable Logic Controllers	2
	Behavioral/Social Science Elective	3
	Communications Elective	3
	Semester Totals	**13**
Fifth Semester		
ELMT	Co-op Electromechanical Technology*	3
	Semester Totals	**3**
	Program Totals	**61**
	Exit Point	**AAS in WET**

* Capstone Course   Course list current as of 1/1/09

As a WET student, you'll not only learn to operate and maintain the electrical, pneumatic, communications, computer, control and/or hydraulic systems that make a wind turbine function, you'll be part of the team that operates and maintains its own operating wind turbine. The TSTC West Texas turbine has the first, full-scale, 60 cycle, 2 MW turbine built by the DeWind Corporation. It was constructed just west of Sweetwater in January, 2008, went online in late February, was approved for connection to the grid on September 1, 2008, and was put into full production a month later.

	Texas State Technical College	
Students Starting Fall 2010	Wind Energy Technology Certificate 2	Total Credits: 49
First Semester		
LEAD	Corporate & Community Development with Critical Thinking	2
WIND	Introduction to Wind Energy	3
CETT	Electricity Principles	4
MATH	College Algebra	3
	Semester Totals	**12**
Second Semester		
WIND	Wind Turbine Materials and Electromechanical Equipment	3
CETT	DC-AC Circuits	4
ELMT	Basic Fluid Power (Hydraulics and Pneumatics)	3
POFI	Computer Applications I	3
	Semester Totals	**13**
Third Semester		
WIND	Wind Business	3
WIND	Wind Power Delivery System	4
INMT	Industrial Automation	4

ITNW	Network +	3
	Semester Totals	**14**
Fourth Semester		
WIND	Turbine Troubleshooting and Repair	4
ELMT	Programable Logic Controllers	2
ELMT	Electromechanical Technology*	3
POFT	Job Search Skills	1
	Semester Totals	**10**
	Program Totals	**49**
	Exit Point	**Certificate in Wind Energy Technician—Certificate 2**

* Capstone Course	Course list current as of 1/1/09

The WET coursework at TSTC consists of important principles and applications of DC and AC circuits, wind turbine construction materials and electromechanical equipment operation and maintenance. You'll learn about the digital fundamentals of industrial automation and PLCs, basics of reading manuals, blueprints and schematic diagrams hydraulics and pneumatics, wind turbine troubleshooting and repair, and of course, the business of wind energy.

Since much of the work of operating and maintaining a wind turbine requires working in different weather conditions at heights of up to 300 feet, your TSTC instructors will emphasize the safety aspects of working in the industry. You'll learn about Supervisory Control and Data Acquisition (SCADA), the computerized standard computerized system that coordinates and controls the wind tower network and its connections to the regional power grid. And you'll discover that the systems you'll master through the WET program are transferrable to many other types of utility companies, providing you with a number of interesting and rewarding career options.

The WET program at TSTC West Texas offers students three programs. One is a twenty-month program culminating in an associate of applied science degree. The AAS curriculum consists of a five-semester track, requiring two years (including one summer) for full-time students to complete. Courses are heavily lab-oriented, providing you with ample hands-on experience and training, led by an instructor who has spent time in the industry.

By the fourth or fifth semester, successful students in TSTC's cutting-edge technology programs typically have jobs waiting for them when they graduate.

While the two-year AAS program requires completion of sixty-one credit hours (almost equally split between lab and classroom hours), TSTC also offers two certificate programs. The Wind Energy Technician Certificate 2 curriculum is essentially a four-semester, sixteen-month program consisting of forty-nine credit hours. This program's primary difference from the two-year degree study is that there are fewer computer, networking and general education courses. The third program is a one-year course of study consisting of twenty-six credit hours over eight months resulting in the awarding of the Wind Energy Technician Certificate 1.

Sample Course Syllabus

The following summary of the syllabus for the TSTC West Texas course WIND 2310 (Wind Turbine Materials and Electromechanical Equipment) provides a glimpse of the variety of subjects covered and the knowledge, skills and tools you will learn as a wind energy technician.

With the prerequisite of the program's Introduction to Wind Energy, the objective of the WIND 2310 course is identification and analysis of the components and systems of a wind turbine. At the conclusion of the course, the successful student will be able to:

1. Describe the potential effects of heat generation on various materials and heat control mechanisms;
2. Define the effects of machining and heat treatment of metals as they relate to predictable failures;
3. Identify gel coats, UV characteristics, flexibility, impact resistance of different coating types and how they are applied;
4. Identify types and specifications of various fasteners and how they are affected by torque, lubricants, hydraulic bolt stretchers, and tensioners;
5. Inspect gears, scaling, types of gear boxes (hybrid, planetary versus helical/parallel shaft), and probable causes of failure;
6. Identify rpm, gear ratios, and failure mechanisms;
7. Identify types, application and compatibility of different lubricants;

8. Identify electrical control system components such as circuit protection devices, sensors, relays, contactors, actuators, timers, counters, motors, and various types of DC and AC drives;

9. Convert units of measurement between metric and U.S. standard; and

10. Demonstrate safety procedures required by OSHA 1910, NFPA, IEEE 519, International Electric Code and National Electric Code Standards.

Other competencies/skills students will study and practice in completion of the requirements of WIND 2310 include allocating time and resources; interpersonal skills like working on teams, cross-training, customer service, leadership; and working with different ethnicities or cultures. Students will acquire skills in collecting, organizing, maintaining and evaluating data by computer and manually, selecting tools and procedures for performing specific tasks, and understanding and working effectively within social, technological, and organizational systems. Emphasis is placed on basic skills like reading, writing, basic math, speaking and listening, as well as thinking creatively, problem solving, making decisions, and learning how to learn and reason.

Each one of the twenty courses in the curriculum leading to the AAS degree is focused on this sort of specific classroom and laboratory learning objective. Every aspect of the training needed to operate, maintain, troubleshoot and repair wind turbines and turbine components is covered in detail in the classroom, but more importantly, in intensive hands-on laboratory activities. As they progress through the semesters, students very quickly begin to build on these introductory knowledge and skills. Fulltime students advance to the associate degree in about twenty months or to one of the certificates within eight to sixteen months.

Although TSTC's WET program provides the education and training for entry-level and advanced placement in the industry, many of the classes in the program are designed to prepare graduating students for alternative careers in technical fields other than wind energy. The wide-open opportunities for employment in wind energy are attracting record numbers of enrollees into the TSTC program. Some drop out early, however, when they begin spending three days a week in the field, accompanying current technicians as they climb turbines, and discover what the job actually requires. It takes a self-confident, motivated individual to feel comfortable climbing and working atop a 260-foot steel tower on a daily basis.

A supremely important benefit for students in the TSTC WET program is the opportunity to network with established industry professionals through the program's advisory committee. The advisory committee is made up of representatives from Texas Tech University, Florida Power and Light, WTX Wind Energy Consortium and GE Wind Energy, among others. As a group, they not

	Iowa Lakes Community College	
Students Starting Fall 2010	Wind Energy & Turbine Technology	Total Credits: 26
First Semester		
CSC-110	Intro to Computers/Information Systems	3
MAT-101	Intermediate Algebra or	3
MAT-102	Intermediate Algebra (4 credits)	
WTT-102	Intro to Wind Energy	3
WTT-114	Field Training and Project Operations	5
WTT-113	Direct Current Electrical Theory	4
WTT-123	Alternating Current Electrical Theory I	4
	Semester Totals	**22**
Second Semester		
BUS-121	Business Communications	3

BUS-161	Human Relations	3
WTT-126	Basic Hydraulics	3
WTT-133	Wind Turbine Mechanical Systems	3
WTT-134	Electric Motors and Generators	4
WTT-244	Alternating Current Electrical Theory II	4
	Semester Totals	**20**
Thrid Semester		
WTT-201	Wind Turbine Site Construction and Locations	1
WTT-932	Wind Turbine Internship	5
	Semester Totals	**6**
Fourth Semester		
PHS-166	Science (Meteorology, Weather & Climate OR Chemistry OR Physics)	4
WTT-214	Basic Networking and Computer Technology	3
WTT-216	Power Generation and Transmission	3
WTT-223	Airfoils and Composite Repair	3
WTT-245	Electrical Practical Applications	4
	Semester Totals	**17**
Fifth Semester		
BUS-121	Principles of Management	3
BUS-161	Wind Turbine Siting	3
WTT-126	Data Acquisition and Assessment	3
WTT-133	Programmable Logic Control Systems	4
	Semester Totals	**13**
Sixth Semester		
WET 290	Optional Internship in WET	6
	Semester Totals	**6**
	Diploma Option, Total Credits	**48**
	Associate of Applied Science Degree, Total Credits	**80**

only help TSTC to identify the knowledge and skills required for entry-level technicians (as mandated by the Texas Higher Education Coordinating Board) but serve as a source of information about, and contacts with, industry employers from all across the nation. Students completing their technical training and education at this level of quality are often pleasantly surprised to realize they have positioned themselves to easily continue their education with an engineering or business degree. They might even decide to attend Texas Tech University in Lubbock, which has a doctorate program in wind science.

For those who wish to go directly into this growing industry, however, graduates will find plenty of opportunities with employers who are almost desperate to find well-trained, experienced workers who understand the

	North American Wind Research and Traning Center at Mesalands Community College	
Students Starting Fall 2010	WET Two Year Program	Total Credits: 26
First Semester		
ACS 100	Student College Success	3
AHS 118R	Adult CPR/First Aid	1
WET 101	Introduction to Wind Energy	2
WET 105	Electrical Theory I	4
WET 107	Workplace Fitness I	1
WTT 115	Field Safety and Experience	4
WET 140	Wind Turbine Climbing and Safety I	3
	Semester Total	**18**
Second Semester		
CIS 101	Intro to Computers	4

GEOL 141	Introduction to Environmental Science	4
WET 204	Introduction to Hydraulics	3
WET 117	Workplace Fitness II	1
WET 121	Wind Turbine Mechanical Systems	3
WET 205	Electrical Theory II	4
WET 141	Wind Turbine Climbing and Safety II	3
	Semester Totals	**22**
Third Semester		
ENG 102	English Composition	3
MATH 107	Intermediate Algebra	3
WET 110	Wind Turbine Operations and Maintenance	4
WET 116	Introduction to Motors and Generators	3
WET 127	Workplace Fitness III	1
WET 202	Power Generation and Distribution	3
WET 240	Wind Turbine Climbing and Safety III	3
	Semester Totals	**20**
Fourth Semester		
COM 102	Public Speaking	3
ENG 233	Professional and Technical Writing	3
WET 137	Workplace Fitness IV	1
WET 210	Wind Turbine Siting and Construction	3
WET 212	Monitoring and Communication Technology	2
WET 215	Wind Turbine Diagnosis and Repair	3
WET 216	Digital Electronics	3
WET 241	Wind Turbine Climbing and Safety IV	3
	Semester Totals	**21**
	Total Credits	**81**

technology, are willing to do the climbing and are capable of getting the job done with little or no additional training.

Wind Energy & Turbine Technology Two-Year Program: Iowa Lakes Community College (Estherville, IA)

This program was the first two-year wind energy program in the state of Iowa. Since its creation, the college has seen enrollment swell with the dynamic increase in wind turbines and farms in the state. Iowa Lakes Community College continues to graduate highly respected technicians who can install and maintain the latest wind turbines.

The program consists of a two-year course of study, with a diploma or certificate available as an option after the first three terms. Those students who wish to earn the diploma in lieu of an associate of applied science degree prepare for entry-level jobs with training and education in wind turbine construction, maintenance and operation, as well as an additional business communications course.

	Minnesota West Community and Technical College	
Students Starting Fall 2010	WET Degree Program Two-Year Degree	Total Credits: 64
Curriculum		
ELCO 1100	Electrical Circuit Fundamentals AND	3
ELCO 1105	Electrical Circuit Fundamentals Lab OR	3
ELCO 1101	DC Circuits AND	3
ELCO 1106	AC Circuits	4
ELCO 1100	Wind Energy Fundamentals	3
ELWT 1110	Mechanical Systems	3
ELWT 1120	Air Foils, Blades, and Rotors	1
ELWT 1150	Wind Turbines	2
ELWT 1160	Environmental, Health, and Safety Wind Energy	1
ELWT 1170	Wind Energy OSHA Standards	1

ELWT 1180	Wind Generation/Transmission/Distribution	3
ELWT 2110	Turbine Siting and Construction	3
ELWT 2130	Data Acquisition and Communication	3
CSCI 1102	Intro to Microcomputers	3
ELEC 1225	Electric Motors	4
ELEC 2205	Electric Motor Control I	4
ELEC 2230	Programmable Logic Controllers	4
ELEC 1235	Applied Electrical Calculations	2
FLPW 1100	Fluid Power Hydraulic Theory	4
EMS 1112	AHA, CPR Healthcare Provider	1
	General Education:	
SPCH 1101	Intro to Speech	3
ENGL 1101	Composition I	3
NSC1 1100	Issues in the Environment	3
GEOG 1100	Intro to Geography OR	3
GEOG 1101	Intro to Physical Geography	4
	General Education Electives	3-4
	Total Credits	**64**

	Minnesota West Community and Technical College	
Students Starting Fall 2010	Wind Energy Mechanic One-Year Degree	Total Credits: 32
Curriculum		
ELCO 1100	Electrical Circuit Fundamentals AND	3
ELCO 1105	Electrical Circuit Fundamentals Lab OR	3
ELCO 1101	DC Circuits AND	3
ELCO 1106	AC Circuits	3
ELCO 1100	Wind Energy Fundamentals	3
ELWT 1110	Mechanical Systems	3
ELWT 1120	Air Foils, Blades, and Rotors	1

ELWT 1150	Wind Turbines	2
ELWT 1160	Environmental, Health, and Safety Wind Energy	1
ELWT 1170	Wind Energy OSHA Standards	1
ELWT 1180	Wind Generation/Transmission/Distribution	3
CSCI 1102	Intro to Microcomputers	3
ELEC 1235	Applied Electrical Calculations	2
FLPW 1100	Fluid Power Hydraulic Theory	4
EMS 1112	AHA, CPR Healthcare Provider	1
	General Education Electives	2
	Total Credits	**32**

	Minnesota West Community and Technical College	
Students Starting Fall 2010	Windsmith Certificate	Total Credits: 16
Curriculum		
ELCO 1100	Electrical Circuit Fundamentals AND	3
ELCO 1105	Electrical Circuit Fundamentals Lab OR	3
ELCO 1101	DC Circuits AND	3
ELCO 1106	AC Circuits	3
ELCO 1100	Wind Energy Fundamentals	3
ELWT 1170	Wind Energy OSHA Standards	1
ELWT 1180	Wind Generation/Transmission/Distribution	3
ELWT 1104	Basic Digital Circuits	2
FLPW 1100	Fluid Power Hydraulic Theory	4
	Total Credits	**16**

Students working to earn the AAS degree finish a second year with training in diagnosis of turbines, high tech-low voltage, computerized control and monitoring systems, composites and composite repair, data acquisition and team management.

Graduates of the two-year degree program are qualified not only for entry-level positions, but will also be prepared to work as a wind turbine operator or even a supervisor.

WET Two-Year Degree Program: North American Wind Research and Training Center at Mesalands Community College (Tucumcari, New Mexico)

The arid areas of western New Mexico are already experiencing a shortage of qualified wind energy technicians. NAWRTC at Mesalands Community College in Tucumcari offers a top quality two-year program in turbine technology and maintenance, tower safety and the business of wind energy. Completion of the first year's coursework earns the graduate an Applied Science Certificate in WET. Students who earn an associate of applied science degree from Mesalands will be ready to meet the demands of this evolving career field, as well as position themselves for rapid advancement.

	University of Iowa	
Students Starting Fall 2010	Wind Power Management Master of Science	Total Credits: 61-66
56:155	Wind Power Management	3
56:166	Stochastic Modeling	3
56:177	Operations Research	3
56:134	Process Engineering	3
56:162	Quality Control	3
56:178	Digital Systems Simulation	3
58:195	Fundamentals of Wind Turbines	3
53:107	Sustainable Systems	3
53:117	Remote Sensing	3
53:251	Environmental Systems Modeling	3
58:1xx	Advanced Energy Systems Design	3
58:143	Computational Fluid and Thermal Engineering	3
58:255	Multi-scale Modeling	3
58:268	Wind Turbine Internship	3
58:195	Aeropropulsion	3
55:050	Communication Systems	3
55:164	Computer-based Control Systems	3
55:181	Formal Methods in Software Engineering	3
55:160	Control Theory	3
Elective Courses in Wind Power Managment		
22C:144	Database Systems	3
6K:234	Information Knowledge Management	3
6K:176	Managerial Decision Models	3
6K:226	Visual Basic Programming	3
6K:228	Web and Multimedia	3
175:195	Occupational Safety	3
44:127	Environmental Quality: Sci. Tech. & Pol.	3

44:135	Urban Geography	3
12:114	Energy and the Environment	3
56:176	Applied Linear Regression	3
General Education Requirements 18 Credit Hours		

WET Degree Programs: Minnesota West Community and Technical College (Canby, Minnesota)

The WET program at Minnesota West Community College offers students an AAS program in WET, a one-year certificate program in Wind Energy Mechanics, and a one-year Windsmith program.

The two-year WET program offers students an excellent curriculum of lecture and hands-on training to prepare its graduates for success in all aspects of troubleshooting, service, repair and infrastructure operations of wind turbine generators and related equipment.

Minnesota West Community and Technical College offers the Windsmith Certificate as an introduction to the Wind Energy Industry. This course of study is ideal for individuals who are interested in the field and wish to gain a working knowledge of wind energy and the ways it contributes to our current national and global industrial energy challenges.

Master's and Doctoral Programs in Wind Power Management: University of Iowa (Iowa City, IA)

Course Requirements for Master of Science Degree

In addition to fulfilling the basic requirements for all master's degree programs in Industrial Engineering, students pursuing an MS in Industrial Engineering with a focus area in Wind Power Management are required to take additional classes, as follows:
- MS Thesis Program: Nine semester hours from Table 1

- MS Non-Thesis Program: Twelve semester hours from Table 1 and three semester hours from Table 2

Students enrolled in the Wind Power Management doctoral program are expected to meet all IE Graduate Program requirements, as well as to expand his/her background in the field by obtaining additional coursework credits in natural science and energy-related classes. For more details, please contact:

Admission: Tim Rocheleau,
trochele@engineering.uiowa.edu

Academic eligibility: Geb Thomas,
gthomas@engineering.uiowa.edu, or

L.D. Chen, ldchen@engineering.uiowa.edu

Courses: Andrew Kusiak,
ankusiak@engineering.uiowa.edu

Doctoral Program in Wind Science and Engineering (WISE): Texas Tech University (Lubbock, Texas)

The Wind Science and Engineering (WISE) doctoral degree program at the Texas Tech University Graduate School is essentially a multidisciplinary degree. A minimum of eighteen semester hours of coursework is required to qualify, which can be fulfilled with a bachelor's or master's degree in a related field such as engineering, atmospheric sciences, economics or physical sciences.

The doctorate degree itself consists of sixty semester hours or more of graduate studies as well as a mandatory dissertation. The program includes core courses, individual field of emphasis courses and an internship. The foundation work for this new degree program is based on twenty-one semester hours of courses in wind and industrial engineering, natural and atmospheric sciences (geographic

WET Instructor Profile: *Roy Limas*

As a dispatch supervisor for General Electric's Wind Energy Service Center, Roy Limas monitors all turbine sites within a sixty-mile radius of Sweetwater, Texas, a total of 851 turbines. In addition to dispatching technicians to malfunctioning wind turbines, he also orders parts, dispatches parts runners, manages tech hours and updates several reports for upper management.

Job satisfaction ranks high among Limas' reasons for working in the wind energy industry. "I was at the right place at the right time when I applied," he said. "Since I've been employed, there are several reasons why I enjoy my career. This field is in its infancy, and at some point in my career I would like to be a major contributor in research and development."

With a bachelor of science degree in mathematics, that is most likely a reachable goal for Limas, especially since GE, like many large corporations, takes pride in promoting from within. With the growth of the business in its current state, management opportunities are increasingly available. And even though Limas came to the industry with an impressive educational background, continuing on-the-job training is an integral part of working at GE, or any other wind energy company.

"With technology always advancing, the need for learning programs is constantly growing," Limas noted. "Currently, GE sends all technicians to GE training schools for various certifications. I enjoy learning everything I can on our 1.5 MW turbines and look forward to learning about our 2.5 MW turbines."

The Fall 2008 issue of *The Occupational Outlook Quarterly*, a publication of the Bureau of Labor Statistics, notes that wind energy is the fastest growing segment of the renewable energy industry in the United States, creating excellent job opportunities for employees who, like Limas, are teachable and eager to learn new technologies. "With our government getting involved and the current push towards green energy, I believe wind energy technology will continue to improve and make strides towards supplying more energy to our countries and cities."

As a result, Limas sees now as an excellent time for young people to seek wind energy careers. He recommends students invest a few years in school, as the pay off can be exponential.

"During our economic crisis, as a nation, GE wind energy is not only stable but also currently looking for a few good techs," he said.

information systems, wind sciences meteorology, wind storm hazards, and field experiments), economics (statistics and wind business management), numerical analysis, and public policy analysis, as well as an additional fifteen hours pertaining to the student's dissertation research. Students are expected to complete all of the core courses by the end of the first year. For more details, contact:

Andy Swift, Andy.Swift@ttu.edu

Sukanta Basu, Sukanta.Basu@ttu.edu

www.wind.ttu.edu/

Top Employers in the Wind Power Field

U.S. Wind Farms

Horizon Wind Energy
www.horizonwind.com/home/
Austin Field Development Office
100 Congress Avenue, Suite 2000
Austin, TX
(512) 370-5275

Lone Star Wind Farm Operations and Maintenance
299 FM 604
Abilene, TX
(325) 548-9463

In addition to the Lone Star Wind Farm in Shackleford County, Texas, Horizon Wind Energy also operates the following:

- Blue Canyon Wind Farm, near Lawton, Oklahoma
- Twin Groves Wind Farm,
  eastern McLean County, Illinois
- Prairie Star Wind Farm, Mower County, Minnesota
- Elkhorn Valley Wind Farm, Union County, Oregon
- Maple Ridge Wind Farm, Lewis
  County, and Madison Wind Farm,
  Madison County, New York
- Wild Horse Wind Farm,
  Kittitas County, Washington
- Somerset and Meyersdale Wind Power
  Projects, Somerset County, Pennsylvania
- Mill Run Wind Power Projects, Fayette
  County, Pennsylvania
- Top of Iowa Wind Farm, Worth County, Iowa
- Tierras Morenas Wind Farm, Guanacaste
  Province, Costa Rica

At a projected cost of more than $2 billion, four new wind farms are being planned for the state of Indiana, with development headquarters in Indianapolis. Indiana Wind Farms, in White County, will be the largest of the Horizon projects, with as many as 660 turbines spread across 100,000 acres. The project is expected to produce upwards of 1,000 MW per year when fully operational, enough to power 300,000 homes.

Additionally, the firm is planning smaller farms in Randolph and Howard counties and four more in Ohio. It also owns and operates seven wind farms in New York, Colorado, Texas, Oregon, Illinois and Minnesota.

E.ON Climate & Renewables

www.eon.com/en/unternehmen/24161.jsp (global)

www.eon.com/en/karriere/26331.jsp (careers)

893 McDonald Road

Big Spring, TX

E.ON Climate & Renewables opened the first two phases of its wind farm in Roscoe, Texas, in September 2008. As of that date the farm was capable of generating more than 335 MW of electricity. Upon completion of all four phases in mid-2009, Roscoe will rank as one of the world's largest wind farms with 627 wind turbines and a total capacity of 781.5 MW—enough to power more than 250,000 homes.

DeWind, Inc.

www.compositetechcorp.com/windpower.htm

2648 FM 407, Suite 200

Bartonville, TX

(940) 455-7450

DeWind, Inc. designs, and manufactures wind energy turbines and converters for customers all over the western hemisphere. The company is based in Bartonville, Texas, and also has operations in Cuxhaven, Germany.

In October of 2008, it was announced by Composite Technology Corporation that its subsidiary, DeWind Inc., is partnering with Higher Perpetual Energy to develop four wind farms in Texas totaling up to 620 megawatts of electrical energy capacity. The partnership will develop four wind energy projects in the Texas panhandle with the potential of generating more than 600 MW. The working wind turbine operated by the students and staff of Texas State Technical College West Texas was made by DeWind.

GE Wind Energy

www.ge.com/careers/
801 E. Broadway
Sweetwater, TX
(325) 794-5100

For nearly twenty-five years, GE has been one of the world's leading wind turbine suppliers, with more than 10,000 wind turbine installations worldwide generating more than 15,000 MW of capacity. GE builds and assembles wind energy turbines and components in Germany, Spain, China, Canada and the United States, and also provides development, operation and maintenance support.

Garrad Hassan American, Inc.

www.garradhassan.com/careers/
1601 Rio Grande Street
Suite 400
Austin, TX
(512) 469-6096

5040 Quail Terrace
Abilene, TX

Garrad Hassan describes itself as a "World Renewable
Energy Total Access Marketing Partner," providing
wind turbine design, testing and certification;
strategic and public policy studies; technical advisory
services for investors; wind farm development
worldwide and many other developmental and
support services. With headquarters in Australia,
Canada, China, Denmark, France, Germany, Italy,
Japan, Mexico, the Netherlands, New Zealand,
Portugal, Spain, Turkey, the United Kingdom and the
United States, Garrad Hassan represents the global
business scale—and the career potential—of WET in
the 21st century. It is also an important partner in
the development of wind energy generation in Texas
and the upper Midwest.

UpWind Solutions, Inc.

www.upwindsolutions.com/home.html
3555 Lear Way
Medford, OR
(541) 608-0755

307 East 7th St.
Iraan, TX

UpWind Solutions provides full-service operations
and maintenance for utility-scale wind projects. The
company also provides comprehensive safety and
EHS training, and specialized technical services.

Additional Wind Energy Systems Businesses in Texas

Amerasian Corporation

6012 Paper Shell Way

Fort Worth, TX

(817) 840-9225

FAX: 817 847 7231

www.amerasiancorp.com

Wind systems design, construction, engineering, project development services

Barr Fabrication, LLC

4501 Danhil Dr.

Brownwood, TX

(325) 643-2277

www.barrfabrication.com

Manufacturer of large wind energy towers and structures and wind energy system components

Cherokee Diversified Services, Inc.

8318 ГM 2728

Terrell, TX

(972) 524-4339

Large and small wind power systems design, installation, construction, project development services, site survey and assessment services, contractor

Cielo Wind Power

823 Congress Ave., 5th Floor

Austin, TX

(512) 615-9463

www.cielowind.com

Design, construction and operation of wind power plants

DCS
14802 Venture Dr.
Dallas, TX
(214) 239-2025
datacentersys.com
Designs, manufactures and installs custom wind energy system components

Hallmark Industrial Supply
5123 Gulfton St.
Houston, TX
(713) 664-7890
hallmarkhose@yahoo.com
Wholesale supplier, importer of large and small wind energy towers and structures

Hill & Wilkinson, LTD
800 Klein Road, Ste. 100
Plano, TX
www.hill-wilkinson.com
Provides consulting, design, installation, construction, engineering, project development services, architectural design services and contractor services for new wind tower installations in Texas and Oklahoma.

Intelligent Electric
2320 Oakmont Dr.
Bedford, TX
(682) 557-0591
Large and small wind energy turbines, towers and structures, and system; installation, construction, site survey and assessment services, maintenance and repair services, testing services

Johnson Plate and Tower Fabrication

PO Box 608

201 Los Mochis

Canutillo, TX

(915) 877-2300

Manufacturer of large steel tubular wind towers

Lowest Cost Energy

9442 CR 254, No. 28

Clyde, TX

(817) 532-7096

Sells, exports large and small wind turbines, small wind energy systems and system components

NGP Power Corporation

125 E. John Carpenter Freeway, Suite 670

Irving, TX

(972) 409-9980

Nicholson Sales

34 East Lakeshore Dr.

Ransom Canyon, TX

1-806-829-2178 or 1-806-829-2179

Supplier of large wind energy system components

RGV-Solar Inc

PO Box 4731

Edinburg, TX

(956) 534-6231

rgv-solar.com

Design, installation and construction of large wind energy system components; engineering, project development services, contractor services

Round Rock Geosciences LLC

1505 Laurel Oak Loop
Round Rock, TX
(512) 496-8728
www.roundrockgeo.com
Geophysical Services for all sizes of wind farms;
consulting, engineering, site survey and assessment
services as well as contractor services

Run Energy, LP

7500 San Felipe
Suite 150
Houston, TX
(713) 458-1550
Provider of large wind turbine operations,
maintenance and construction; construction,
contractor services, maintenance and repair services,
testing services

Superior Renewable Energy, LLC

1600 Smith, Ste. 4240
Houston, TX
(713) 571-8900
www.superiorrenewable.com
Provides wind power development and construction
services

Texas State Technical College West Texas

300 College Dr.
Sweetwater, TX
(325) 236-8283
www.westtexas.tstc.edu/
Large wind energy systems, turbines, wind energy
towers and structures' education and training
services

Trinity Structural Towers, Inc.

2315 North Main St.
Suite 110
Fort Worth, TX
Manufacturer of large wind energy system
components, wind energy towers and structures

Additional Wind Energy Systems Businesses Outside of Texas

Arkansas

Custom Home & Commercial Electronics
dba Coger Custom Construction
2109 W. Loren Circle
Fayetteville, AR
(479) 841-7116
www.ch-ce.com
Sales, installation, service, construction services, small wind turbines

Green Energy Products
1 Main St.
Ponca, AR
(870) 861-5252
Retail sales of small horizontal and vertical axis wind turbines

Arkansas Solar Power
2771 Old Farmington Rd.
Fayetteville, AR
(501) 442-7652
Provides alternative energy education, training, and consultation; sales and service of solar, wind and hydropower components

Energy Resources
c/o 4700 McCain Blvd., #15954
North Little Rock, AR
(501) 758-4444
Retail sales of small wind energy system components

Rocky Grove Sun Company
3299 Madison 3605
Kingston, AR
(479) 665-2457
Retail sales, design and installation of grid-tied PV and wind systems as well as wind generators

Winds of Change

5801 Hwy 94 N
Rogers, AR
(479) 616-8077
Retail sales, consulting, installation, project
development services, site survey and assessment
services on small wind energy systems; wind energy
assessment equipment

Colorado

Rocky Mountain Wind Systems

885 West 9th Ave.
Broomfield, CO
(303) 464-8323
Wind turbine broker

Toltec Energy

938 Quail St., Unit A
Lakewood, CO
(303) 202-0150
Renewable energy engineering, research, testing
services, and wind power plants

Spider Energies

871 Thornton Pkwy, Ste. 129
Thornton, CO
(248) 770-6395
Manufacturer, wholesale supplier of small wind
turbines

Louisiana

Beaird Industries, Inc.

601 Benton Kelly St.
Shreveport, LA
(318) 865-6351
Engineering and manufacturing of large wind
energy towers and structures

Effective Solar Products, LLC

PO Box 269

Mathews, LA

(504) 537-0090

Retail sales, wholesale supplier of small wind energy systems

New Mexico

AAA Solar Supply

2021 Zearing NW

Albuquerque, NM

(800) 245-0311

Retail sales, wholesale supplier of small wind powered electric generators

All Star Electric

10000 Trumbull SE, Suite F

Albuquerque, NM

(505) 856-1010

Design, engineering, project development, construction and engineering of large and small wind energy systems

Energy Concepts

HC 68, Box 11AA

Sapello, NM

(505) 454-0614

Consulting, design, installation, contractor services, maintenance and repair for wind energy systems

Geophysical Phased Energy (GPE) Technologies Corporation

4020 Vassar Dr. NE, Ste. H

Albuquerque, NM

(505) 350-5636

Provides wind, solar and water power generation systems

Solar de Taos
P.O. Box 159
Carson, NM
(505) 751-0620
Retail sales of small and medium wind energy
systems

Solmomma, Inc.
PO Box 1980
Moriarty, NM
(505) 832-2722
Retail sales, exporter of wind energy systems,
turbines, components, towers and structures

Southwest Building & Energy Technologies
625 South Roosevelt Rd. 11
Portales, NM
(806) 787-0298
Installation, construction of large and small wind
turbines

Taos Wind Power, Inc.
212 Paseo Del Pueblo Norte
Taos, NM
(575) 751-9463
Project development services, research, site survey
and assessment services for large wind energy
systems, towers and structures

Oklahoma

Suspended Climbing Systems
6998 S. 145th E. Ave.
Broken Arrow, OK
(918) 251-2011
www.suspendedclimbingsystems.com
Manufacturer of a chain-driven powered platform
for use in large wind turbine towers

Bergey Windpower Co.
2200 Industrial Blvd.
Norman, OK
(405) 364-4212
Small wind powered electric generators

Harvest Solar and Wind Power
1571 East 22 Place
Tulsa, OK
(918) 743-2299
Small wind energy systems

Solarwellpumps.com
Route 1 Box 52A
Balko, OK
(866) 483-6851
Small wind energy system components and turbines

Wind Energy Technician Recruiters

Careers in Wind
www.careersinwind.com/

Energy Placement
www.energyplacement.
com/jobs/renewable_energy_jobs_alternative

Global Windpower Services
www.global-windpower.com/jobs.php

GreenJobs.com
www.greenjobs.com/public/index.aspx

iHireEnvironmental.com
www.iHireEnvironmental.com

Indeed.com Wind Energy Jobs Search Engine
www.indeed.com/jobs?bloc=1&q=wind+energy

NationJobs.com
www.nationjobs.com

RenewableEnergyJobs.net
www.renewableenergyjobs.net/

WindEnergyJobs.com
www.WindIndustryJobs.com

U.S. Wind Energy Industry Associations

The American Wind Energy Association (AWEA)
www.awea.org
AWEA events and links to other wind and renewable energy events and conference calendars. Lists wind power plants and development projects throughout the U.S.

The California Wind Energy Collaborative
cwec.ucdavis.edu/activities/
A partnership of the University of California and the California Wind Energy Commission; publications and other information about wind energy in California

National Wind Coordinating Committee
www.nationalwind.org/
Identifies environmental issues affecting the wind power industry

Windustry (Minnesota)
www.windustry.org/
Wind energy basics, business opportunities in wind energy, resource library, news and events, legislation

WindPowering America
www.windpoweringamerica.gov/
American government agency for the promotion of the use of wind power; regional activities, calendar, and online publications

Women of Wind Energy
wowe@windustry.org
WoWE promotes the engagement, professional development and advancement of women in the wind industry

Utility Wind Interest Group
www.uwig.org/
Promotion of wind energy for utility applications; workshops, wind resource assessment, etc.

International Associations

The international listings below link only to English-language sites. There are numerous additional wind energy websites based in Austria, Denmark, Germany, Estonia, Finland, France, Norway, Sweden, Poland, and elsewhere that can also be found on the Internet.

African Wind Energy Association
www.afriwea.org
A non-profit organization to promote and support wind energy development on the continent of Africa

Australian Wind Energy Association
www.auswea.com.au/
Industry hub for the Australian wind energy community

The British Wind Energy Association
www.britishwindenergy.co.uk/
Wind energy industry source in the United Kingdom, the trade and professional association

Canadian Wind Energy Association
www.canwea.ca/
Wind energy production industry in Canada

Collaborative Offshore Wind Research into the Environment (COWRIE-UK)
www.offshorewindfarms.co.uk/Pages/COWRIE/
Web site for the offshore wind industry in the United Kingdom

Danish Wind Turbine Manufacturers' Association
www.windpower.org/en/core.htm
Wind energy facts, news, history, wind maps, wind farms in Denmark

European Wind Energy Association
www.ewea.org/
Promotion of wind energy development in the EU; overview of European wind farm capacity, publications

Indian Wind Energy Association (InWEA)
www.inwea.org/
Promotes the wind energy to Indian industry and government, policy makers and consumers

Indian Wind Turbine Manufacturers Association (India)
www.indianwindpower.com/
Wind industry in India; wind development site requisites, wind potential, state approval, financial benefits

Irish Wind Energy Association
www.iwea.com/
Promotion of wind energy in Ireland

Latin America Wind Energy Association
www.lawea.org/ing
Promotion and advocacy of wind energy in Latin America

Les Compagnons d'Eole (Belgium)
users.swing.be/compagnons-eole/windturbine/eole_us.htm
Promotes the wind power industry in Belgium; library, quarterly newspaper

New Zealand Wind Energy Association
www.windenergy.org.nz/
Central site for the advocacy of wind energy in New Zealand

Offshore Wind Energy Network (UK)
www.owen.eru.rl.ac.uk/
Promotes research and development of the UK's offshore wind resource

Sahara Wind
www.saharawind.com
Information on West African Sahara wind energy potential and applications; often rated one of the best wind power sites in the world

WindPowerIndia
www.windpowerindia.com/
Publications, directory, statistics dedicated to the wind power industry in India.

World Wind Energy Association
www.wwindea.org/
Global conferences, exhibitions, and publications

Wind Energy Industry Publications

American Wind Energy Association (AWEA)
www.awea.org/

Alternative Energy News
www.alternative-energy-news.info/headlines/wind/

GreenBiz.com
www.greenbiz.com/

National Renewable Energy Laboratory
www.nrel.gov

National Wind
www.nationalwind.com/

Renewable Energy News
energy.sourceguides.com/news.shtml
Sustainable Energy Coalition
www.sustainableenergycoalition.org

U.S. Dept. of Energy-Energy Efficiency and Renewable Energy
www1.eere.energy.gov/education/index.html

Wind Energy Weekly (AWEA, by subscription)
www.aweastore.com/
Detailed and up-to-date information on wind energy research and development.

Windletter (AWEA, electronic publication, by subscription)
www.awea.org/membercenter/

Windletter Archives
www.awea.org/windletter/windletter_archive.html

Wind Power Handbook
www.windpowerhandbook.com/

Wind Power Monthly
www.windpower-monthly.com

Additional Suggested Resources

AGORES
www.agores.org/SECTORS/WIN_TWI/default.htm
A global overview of renewable energy sources

AWEA Resources Online Library
www.awea.org/resources/resource_library/
Documents and reports, fact sheets, databases, A/V materials, archives, renewable energy blogs, and more resources for wind energy professionals

Global Wind Energy Council (GWEC)
www.gwec.net/
A collection of wind energy articles, events, and activities worldwide

National Sustainable Agriculture Information Service (ATTRA)
attra.ncat.org/farm_energy/wind.html
Information and links to articles about wind energy and agriculture

National Wind Coordinating Collaborative Online Publications Links

www.nationalwind.org/publications/

Links to publications on wildlife/wind interaction, wind power distribution, papers on wind energy issues, state policies on turbine siting and permits, and more

National Wind Technology Center

www.nrel.gov/wind/publications.html

The center's library provides a list of recent wind energy publications and a link to the searchable publications database at the National Renewable Energy Laboratory containing publications dating back to 1977

Wind Powering America Publications Online Database

www.windpoweringamerica.gov/publications.asp

More than 280 publications dating back to 1994 relating to all aspects of the wind power industry

Wind Power Online

www.windpoweronline.com/

A worldwide portal for the wind power industry with articles about the industry in Spain, France, Germany, Italy and more

INDEX

A

B

E

R

S

About the Author

Mike Jones

Mike Jones is a freelance writer based in Waco, Texas. He is a transplant from New Mexico where he graduated from the University of New Mexico in Albuquerque with training in writing for theater and broadcast media. He worked extensively in the broadcasting, advertising, and marketing fields prior to relocating to Texas to work as a writer/producer of instructional and student recruitment videos for Texas State Technical College. More recently, he has been involved in technical instructional curriculum research and development, as well as freelance media writing and production.

Established in 2004, TSTC Publishing
is a provider of high-end technical
instructional materials and related
information to institutions of higher
education and private industry. "High
end" refers simultaneously to the
information delivered, the various delivery
formats of that information, and the
marketing of materials produced. More
information about the products and
services offered by TSTC Publishing may
be found at its Web site:
http://publishing.tstc.edu/.